Co-innovation Dynamics

Smart Innovation Set

coordinated by
Dimitri Uzunidis

Volume 20

Co-innovation Dynamics

The Management of Client–Supplier Interactions for Open Innovation

Romaric Servajean-Hilst

iSTE

WILEY

First published 2019 in Great Britain and the United States by ISTE Ltd and John Wiley & Sons, Inc.

ISTE Ltd
27-37 St George's Road
London SW19 4EU
UK

www.iste.co.uk

John Wiley & Sons, Inc.
111 River Street
Hoboken, NJ 07030
USA

www.wiley.com

Library of Congress Control Number: 2019931459

British Library Cataloguing-in-Publication Data
A CIP record for this book is available from the British Library
ISBN 978-1-78630-331-8

MIX
Paper from responsible sources
FSC® C013604
FSC www.fsc.org

Contents

Foreword

Henry Chesbrough coined the term Open Innovation in 2003, and since then practitioners and academics alike have embraced the concept. He defined Open Innovation as "*the use of purposive inflows and outflows of knowledge to accelerate internal innovation, and expand the markets for external use of innovation, respectively*" [CHE 03]. This definition was inspired by his experience in large companies such as Xerox, IBM, and P&G. Open Innovation has later on been applied to firms in service industries, small companies and start-ups, governments, non-profit organizations, etc. Furthermore, Open Innovation has not only been studied at the firm level: new insights about its management have been developed by taking R&D projects, Open Innovation managers, single relationships with partners, and events such as innovation contests, as units of observations. Similarly, we get an enriched understanding of Open Innovation from new studies that examine the concept at the level of interorganizational networks or ecosystems, and by the integration of Open Innovation into the regional innovation systems framework. Consequently, Open Innovation is gradually moving to the center of innovation management research. However, this does not mean that there are no white spaces in this field: there are many, and managing collaborative innovation between suppliers and clients is one topic that hasn't been analyzed in detail.

In this book, Romaric Servajean-Hilst focuses on the benefits and management challenges when suppliers and clients participate in joint innovation projects. It is a fascinating book that goes into great details as to how to manage such joint initiatives. Suppliers and clients face great difficulties when it comes to managing innovation. They seek to understand the right levers for action and there are many difficulties, because collaborative innovation between client and supplier is prone to a number of specific challenges that are not present when a company is working, for instance with technology partners such as universities or university labs.

Let me explain this in more detail. In any innovation project, there is uncertainty. This uncertainty increases when the resources and results are shared, as is the case when clients and suppliers collaborate. Years ago, I was invited by a leading FMCG (Fast Moving Customer Goods) company to have a look at the management of the company's joint innovation with its main strategic suppliers. The arrangements and contracts with its suppliers were longwinded, reflecting the complexity of the relationship when a major client is innovating with strategic suppliers. The relationship is complex because on the one hand, suppliers and clients have standard contractual arrangements determining the prices and quality criteria for components and raw materials that they already trade. These negotiations between suppliers and clients can be tough because it is a zero-sum game. On the other hand, they start collaborative innovation where the logic is completely different: by relying on each other's strengths and assets, they can create highly novel products. This is essentially a positive sum game and the joint value creation is a function of their openness and teamwork in the innovation process. Yet, at the same time, they have to divide the benefits of the collaboration in a way that does not jeopardize the good teamwork, which guarantees the results of the value creation process. So, they have to finesse an optimum level of competition and collaboration – competition for maximum value capturing and collaboration for superior value creation [ADN 17, HAN 18, KAP 13]. This tension is not apparent in Open Innovation deals with universities or research laboratories that usually operate as a research contractor or license the technology: there is no further collaboration in the product(ion) space as in the case when suppliers and clients co-innovate. The relationship is also more complex than in the case of collaboration with competitors because competitors can make orderly agreements about the way they will commercialize the technology once they finish the joint innovation. Similarly, collaboration with companies from other industries is fairly easy to handle because partners each have their own industry and further interfering in their product markets is almost by definition excluded.

So, what are some difficulties you can expect when suppliers and client companies are innovating together? First, both parties will have to agree about Intellectual Property Rights. Deals regarding Intellectual Property are possible as the supplier may be more interested in the purely scientific or technological Intellectual Property, while the client firm will have stronger appetite for Intellectual Property that relates to the application of that technology. Second, there should be a clear understanding of the limitations of a possible solution: the client firm will usually come up with a cost-restriction for the innovation – for instance a laundry detergent cannot increase more than 5 cents per kg. Third, risk management should be dealt with appropriately: the risk balance is asymmetric, as the supplier has to do most of the initial R&D investments. If the client firm finally decides not to commercialize the innovative product, then it has to foresee this possibility in a contract with the supplier and compensate the latter in one way or another. Lastly, there is genuine tension between both firms because the client firm wants to have broad exclusivity with respect to the innovative product while the supplier has a strong drive to limit this exclusivity and sell the component to the client's competitors.

This example is just an illustration to underline how important it is to understand the complex relationship between suppliers and client firms when they co-innovate. It is a topic that has, in error, often been left aside by Open Innovation scholars. *Co-innovation Dynamics*, written by Romaric Servajean-Hilst, is thus more than welcome and is important to better understand how co-innovation in client-supplier relationships can be managed properly. Here are a few takeaways from the book: first, co-innovation in client-supplier relationships requires a dynamic approach in which the relationship evolves, from its beginning, through a continuous cycle of achievements, evaluations and adjustments. Second, there are four main management methods adopted by client-suppliers that engage with joint innovation projects. Finally, there is a major role for the purchasing department in managing this relationship.

Enjoy the reading!

Prof. Dr. Wim VANHAVERBEKE
Professor of Strategy & Innovation Management at Hasselt University
Visiting Professor at ESADE Business School
Visiting Professor at the National University of Singapore

Acknowledgements

This book presents the results of PhD work conducted in the Management Research Center of Ecole polytechnique, France, under the direction of Prof. Sihem Ben Mahmoud Jouini.

It received the support of Ecole polytechnique foundation, of DELCAR and of innov'&.

Special thanks go to Hubert and Maud without whom this book would not have been what it is today.

Introduction

In 2007, Asseat, a firm pertaining to the automotive sector, approached the Captor firm in order to apply one of its sensors to the automotive world. In order to determine which of the sensors would be suitable, and according to which specifications, tests were carried out over a period of 2 years on the basis of prototypes that were provided by Captor. Once the sensor was identified, a contract was signed. It stated that each party would cover its development costs, each party would keep the intellectual property with in its main field of activity, and in the event of success, Asseat would entrust Captor with the product supply that was developed for the first series. Two years later, many technical difficulties were solved and the product was finalized. Another two years later, Captor supplied Asseat with the developed product set, and a new production line was opened on the other side of the world.

Certifname, a start-up created in 2007, developed a system for verifying the digital identity of Internet users. In mid-2009, Numtel contacted it in order to develop a collaborative certification project that combined virtual identity and real identity. Nine months later, discussions began on the cooperation agreement that would be signed in the first quarter of 2011, when the first version of the experimental platform was already completed. Technical problems then began to arise. Six months later, Numtel terminated the cooperation. One month after that, Numtel launched a product with the same functionalities as the one covered by the cooperation agreement, including some common technological elements. Certifname then sued Numtel for the infringement of intellectual property rights and unlawful exclusion. According to Numtel, which defended itself against such accusations, the product technologies were different.

Ingrecolor is a reference firm in the world of pigments. In 2009, it proposed to its client, Cosmect, the development of a new pigment of a with new functions for cosmect's transparent packaging. A contract was signed: Cosmect's committed to purchasing this new pigment from Ingrecolor in exchange for a exclusivity on the cosmetics market. A few months later, the project came to an end, the pigment being successfully developed. Unfortunately, the technical–commercial equation was not solved, and as agreed, the cooperation ended. Ingrecolor then wished to offer its new pigment to other clients in other sectors, but was not able to do so: Cosmect had unilaterally filed a patent covering all the applications targeted by Ingrecolor. Ingrecolor continued to supply Cosmect with its products, but no longer gave it priority access to its innovations.

In 1997, the car manufacturer Simplauto selected its supplier Simplfour to jointly develop a new opening system. This was a very ambitious system for both companies. Unlike the other contracts between Simplauto and Simplfour, an agreement of just a few pages was signed: it was brief and the respective responsibilities were not defined. Simplauto contributed to Simplfour's study costs. Intellectual property rights were distributed as the project progressed. At the end of 2000, the first vehicle was launched. Quality problems were jointly solved and Simplfour supplied the first-series vehicles. Later, the application of the system would be developed with another Simplauto supplier on a second platform. Since then, this system has been increasingly applied in Simplauto's ranges, with several different suppliers. Simplfour, for its part, has expanded its product range by offering a generic and modular version of the system to other manufacturers.

These four real-life stories reflect a fraction of the variety and complexity of the issues encountered in Open Innovation between a client and a supplier.

Regardless of their position, whether they act as a supplier or a client, companies always face great difficulties when it comes to managing innovation cooperation with a client/supplier. All seek to identify and understand the right levers for action, as well as the contexts in which to use them to make them effective. There are many difficulties, because collaborative innovation between the client and the supplier combines both issues related to inter-firm cooperation and innovation [TAK 01].

In any innovation project, there is a portion of uncertainty. This uncertainty is all the greater when the specified degree of innovation is high. It concerns both the results and the means used to achieve them. This uncertainty is increased when these resources and results are shared, as is the case when clients and suppliers collaborate. It then affects both the client and the supplier.

It may be difficult, if not impossible, to predict what will be discovered and implemented, what will be implemented during and after the innovation project, as well as all the impacts and potential benefits of such a collaboration. The transactional approach to the business relationship that would result from a cost–benefit calculation [WIL 93] is not conceivable. It is not possible to assess the cost or benefit of the transaction project upstream, or the transaction cost levels that would directly lead to its abandonment.

This uncertainty leads to an ontological incompleteness of the contracts formed between clients and suppliers when innovation is concerned. In this case, to the great displeasure of lawyers, it is only rarely possible to specify all the actions and results expected by the parties, as well as the means of measuring them. The contractual arrangements also need to be tailored in comparison with "traditional" contracts [GAR 94, MID 96, p. 21].

Other strategies are therefore deployed in order to manage Open Innovation between clients and suppliers. A relational approach will consist of increasing the number of information exchange mechanisms between companies, on technical, commercial and organizational issues. The presence of "*resident engineers*" on the project platforms is an example of this [LAK 06a, MID 98].

These exchanges allow each party to seek and assess information on the competency and credibility of the other in the face of the challenges put forward by innovation, and to propel themselves into the future. They help to reduce the level of risk perceived by everyone and promote trust between them and between their staff. A consistent level of trust allows a welcoming degree of stability for the success of a relationship, within a context that is inherently unstable; that of innovation.

However, internally, there are still difficulties in accommodating the logic and interactions related to innovation with those related to trade in goods and services that are traditionally dissociated in companies [MID 97] and often poorly integrated [NPV 14]. This results in the establishment of separate governance arrangements. First, the entities dedicated to the client–supplier relationship (the logic of exchanging goods and services, purchasing and selling) are separated from the entities dedicated to the innovation relationship (innovation projects). Second, the relationship management mechanisms are differentiated because they use different standards and assessment methods, in addition to not being engaged in the same time frame.

Successfully managing client–supplier Open Innovation therefore requires the ability to orchestrate the client's internal organization, the supplier's internal organization and the terms of the relationship [MID 07, SÄF 14, TAK 01]. It also requires not considering the innovation project as an "island", but as part of a context and dynamic that goes beyond its own limits [ENG 03][1].

This is the focus of our research, the objective of which is to *understand how the client–supplier Open Innovation performance is built over time*, and the results [SER 15] of which are presented in this book.

In order to meet this objective, we followed a research itinerary by constantly discussing theory and practice [DUM 13]. For four years, we have assembled three major research approaches, both qualitative and quantitative, all while maintaining the same unit of analysis: the client–supplier couple cooperating on an innovation project (Figure I.1).

We have sought to multiply and compare points of view in order to draw intelligible representations. Empirical observations guided the use of appropriate theoretical frameworks [GLA 09], as much as they guided our observations.

First, we defined a conceptual model based on interviews with all kinds of stakeholders in client–supplier Open Innovation partnerships, and on the academic literature. Then, we conducted a quantitative study from the perspective of 160 supplier firms that had at least one innovation project in cooperation with one of their clients. Simultaneously, we conducted a

1 The title of Engwall's article [ENG 2003] "No project is an island: linking projects to history and context" was inspired by a 16th-Century poem by John Donne, "No man is an island entire of itself; every man is a piece of the continent; a part of the main", but maybe also by the article "No business is an island: The network concept of business strategy" by [HÅK 89].

qualitative study in immersion for two years in the life of a client firm – so as to live and follow an innovation project with a supplier.

		Step 1: design of a conceptual model	Step 2: quantitative study	Step 3: qualitative study inspired by ethnography
Research methods	Data collection	• 35 semi-open interviews • Literature review	• Survey by questionnaire • 160 participants	• 2-year observant participation • Unique case study
	Analysis	• Abduction • Multi-thematic coding	• Hierarchical ascendant classification and principal component analysis • Khi-2 test	• Narrative • Visual mapping
Level of analysis	Unit of analysis	• Client-supplier dyad		
	Perspective	• Client, supplier and expert-actor	• 1 actor of the dyad, supplier side	• 1 actor of the dyad, client side
	Level of analysis	• Entities-firms	• Firms	• Entities-firms
Validity criteria	Internal	• Theoretical triangulation • Quantity and variety of data collected	• Use of validated measurement scales • Validation of the survey with practitioners and academics • Quantity and accuracy of data collected	• Length of the observation • Triangulation of sources • Dialogue with stakeholders • Search for rival hypotheses
	External	• Dialogue with practitioners and academics	• Dialogue with practitioners and academics	• Analytical validity • Limited to context
Main limitations		• Theoretical approaches used	• Unique respondent	• Unique case study

Figure I.1. *The three methods of this research itinerary on client–supplier Open Innovation*

This book retraces the results of this research trajectory. In Chapter 1, we present what the academic literature proposes in terms of managing client–supplier Open Innovation and understanding the different sides of its dynamics.

In Chapter 2, we present the combined result of our first two research approaches: the composition, conditions and impacts of the governance of client–supplier Open Innovation, and their links with performance.

Chapter 3 is devoted to a qualitative and longitudinal approach to the dynamics of Open Innovation. With the same interpretation framework of the previous chapter, the 20 months of a client–supplier relationship around an innovation project are reported and then analyzed.

Finally, we conclude this book by presenting the conceptual model of interaction dynamics of client–supplier Open Innovation, the central result of our research, and discuss its contributions. We then highlight the different lessons to be learned from this work for better practice of collaborative innovation between the client and the supplier.

1

Open Innovation, or Collaborative Innovation, Between Clients and Suppliers

At the foundation of Open Innovation, a term forged by academia, is the principle that in a world where knowledge is very widely distributed, one firm alone cannot control it all.

The dominant model of technological innovation practice throughout the 20th Century was that of the R&D laboratory that was vertically integrated into the commercial structure [LEM 06, WES 14]. Innovation was "closed". It was developed from A to Z within the limits of the same firm.

During the second half of the century, innovation gradually opened up; firstly in the client–supplier relationship.

From the 1970s onwards, outsourcing became an increasingly important strategy for companies. The complexity of products increased with the multiplication of subcontracted components. Suppliers played an increasingly important role in new product development projects.

The academic world then began to focus on the subject by studying R&D alliances and technological partnerships[1] as well as the development of new products. The latter initially focused on the automotive sector. Moreover,

---

1 The study of R&D alliances and inter-firm technology partnerships addresses the inter-firm relationship through the pooling of technological resources and the forms that organizations take in order to achieve this pooling. This research work solely focuses on R&D activities [DOZ 87, GUL 95].

from the very first published works [IMA 85, TAK 86][2], it was revealed that the early involvement of suppliers in development projects was one of the keys to the performance of Japanese industry [JOH 09, p. 188]. Among car manufacturers, their involvement thus represents 30% of the total engineering effort, while it represents 16% in Europe and 7% in the USA [CLA 91].

During the 1990s, research extended to sectors other than the automotive sector [EIS 95]. Client–supplier interactions were increasingly recognized as a source of innovation [MID 93, FSS 00]. The question of when to integrate the supplier into the innovation process began to be raised, in addition to how to organize this integration [BID 98, KES 98, MID 97].

However, from the beginning of the 21st Century, studies on the link between supplier involvement in innovation and its performance have yielded contradictory results: they are not necessarily positive, particularly for projects with high technological uncertainty [HOE 05, MEL 14]; the benefits and risks associated with this involvement have been identified [HAN 99].

Research therefore focuses more on the need to adapt the relationship and emphasizes the need to manage both internal and external interfaces [CAL 00, LAK 01, TAK 01, VAN 08]. During the first decade of this century, interest in integrating suppliers into the development of new products has continued to grow, both for researchers and practitioners [SCH 10]. This is also the case for all forms of innovation that go beyond the firm's limits.

Open Innovation, as opposed to the closed innovation of yesteryear, is a term coined by Henry Chesbrough in his 2003 bestselling book "*Open Innovation: The New Imperative for Creating and Profiting from Technology*" [CHE 03].

In a context where innovation is intensifying, opportunities for collaboration are expanding and available resources are becoming scarce, companies no longer have any choice but to open up to external players in order to innovate. They do this through Open Innovation.

---

2 These are two successive versions of the same research work on organizational learning between Japanese companies.

Open Innovation thus groups together the practices that organize and implement innovation activities involving both internal and external parties in a firm. Although it is not a new idea, its development at such a level and its institutionalization is.

The use of the term Open Innovation highlights the intensification, variety and breadth of cooperation opportunities. Academic research on the subject focuses on the study of knowledge flows between organizations (their identification, combination and exploitation) for innovation purposes and related economic models.

Open Innovation between a client and a supplier therefore takes from both the world of Open Innovation and the world of early involvement of suppliers in development projects. These two lines of research are now drawing closer together; this book is a vector of that research.

With that in mind, in this chapter we will present what the academic literature teaches us about the modalities of interactions between clients and suppliers in Open Innovation by following the different episodes of involvement regarding a partner in an innovation project. We will then propose a reviewal of different ways of theoretically understanding the dynamics of inter-firm cooperation within this context.

1.1. Managing client–supplier interactions in Open Innovation

These interactions take place not only through the exchange of money for products and services, as is the basis of client–supplier relationships, but also exchanges of information and knowledge, as well as social exchanges [HÅK 82][3].

Many players interact in client–supplier innovation cooperation. These interactions occur through successive episodes of exchanges within each firm, between individuals, between functions, and between each

3 As stated by The IMP, Industrial Marketing and Purchasing (Group) which is a network of researchers that are interested in marketing, purchasing, innovation and management issues from the perspective of interactions in BtoB and BtoC contexts. Since the mid 1980s, it has provided conceptual models for understanding the mechanisms of continuous interactions and interdependencies between exchange partners [DON 05]. However, innovation is not a specific subject of study in this work.

firm, between their members, on an individual basis or representing their function.

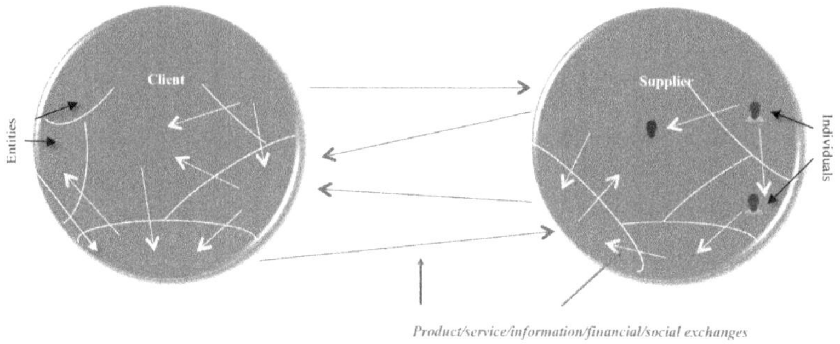

Adapted from Håkansson and IMP Project Group, 1982

Figure 1.1. *Interactions within the client–supplier relationship (adapted from [HÅK 82])*

It is these interactions that bring to life, build or destroy the relationship between a client and a supplier surrounding an innovation project, in which they are jointly involved. In the following, we describe the major episodes that take place across the lifetime of such a relationship and what they involve in terms of management.

1.1.1. *Sourcing Open Innovation selecting a partner and choosing the time of their involvement*

The selection of an appropriate innovation partner is one of the managerial skills that the client must develop in order to ensure the success of innovation client–supplier relationships [CHE 11, p. 223, LED 11, PET 05]. This episode of research and innovation partner selection, or the external knowledge sourcing episode, is the one that receives the most attention from researchers [WES 14].

Operationally, sourcing involves answering two intertwined questions: *when should one involve a supplier in an innovation project? And which supplier should be involved?*

1.1.1.1. *When should a supplier get involved in an innovation project?*

The issue of involving a supplier in an Open Innovation project is trying not to involve it "*ahead of time*". It is a question of knowing when the "*right time*" is [BEN 04, p. 177]. The timing, also related to the choice of supplier, is considered a crucial step in the success of a development project involving a supplier [DEK 08].

The timing of supplier involvement varies according to the uncertainty and complexity of the project, the role to be played by the supplier, and the capacity of potential suppliers available to cooperate on the project [HAN 99, LED 10]. It can be involved at all stages of the innovation project, from the generation of new ideas to production. In addition, the sooner a supplier is involved, the more it will be able to contribute its viewpoint on how to create the project.

In the case of projects with many significant uncertainties and a lack of internal resources to resolve these uncertainties, one solution for the client firm is to work with several potential suppliers (two or three at the most) in a preliminary phase, before choosing the supplier that will be involved [MID 00, p. 11].

Another solution is for the supplier(s) to be consulted in the very early stages of the innovation project, in order to provide expertise on the design and manufacturing methods of the part that could concern it/them [LED 10]. The supplier then acts as a "silent designer"[4] who comes to participate in the innovation project without being integrated into it; the supplier sheds light on the choices of the client's innovation team.

Its importance is not only based on the selection itself, but also on the selection process, which is an opportunity for inter- and intra-firm exchanges.

1.1.1.2. *Which supplier should be involved in the project?*

As a prerequisite, in order to be able to select an external partner, it is necessary to be able to recognize knowledge a capacity for external

4 According to Gorb and Dumas [GOR 87, p. 150], the "silent design" is the "design (realized) by people who are not designers and are not aware that they are participating in design activity".

innovation [COH 90][5]. It is then necessary to be able to recognize how this base of external knowledge (of technologies, markets, etc.), this external innovation capacity, is complementary to the knowledge base of one's own organization.

Many companies have learned this at their own expense when they have sought to apply Open Innovation to the letter, and radically so – forgetting that Open Innovation is not the opposite of closed innovation, but its counterpart. These companies have decided to completely outsource their innovation. They have closed their internal department that was dedicated to innovation. Due to the lack of staff able to correctly identify and qualify external sources of new knowledge, they have no longer been able to generate effective collaborative innovation projects, and in the end, they have lost all innovation capacity.

In addition, the synergy between a client and a supplier for an innovation project is an important determining factor for the future success of the project. It can be summarized into three alignments that can be searched and/or generated:

1) technological alignment. The supplier's ability to respond to client issues, the synergy of their knowledge of markets and technical resources, and the accuracy of their bases of knowledge [EMD 06];

2) strategic alignment. The congruency of their motivations, objectives and strategic visions which are, for example, materialized by the strategic roadmaps of the client and the supplier [HAN 99]. It is also based on the relative position of each firm facing the other [SCH 06], as exemplified by the proposition in the supplier portfolio theory[6] [KRA 83];

3) relational alignment. The compatibility of their culture and their propensity to adapt to each other's processes and changes. It also corresponds to the client's ability to be attractive to the supplier, which particularly implies being able to establish a relationship of trust [ELI 12, SCH 06, SCH 12].

5 This definition relates to the absorptive capacity, which consists of four main stages or abilities which are (1) the recognition of the value of external knowledge, (2) the assimilation of this knowledge, and its application through (3) the transformation and (4) the exploitation of external knowledge absorbed by the internal knowledge base [ZAH 02].

6 And its application through the supplier panel management tools.

These alignments act according to the stage of progress of the innovation project. In addition to ranking potential suppliers, the selection criteria that validate the alignment must be adapted to the timing when the supplier is integrated.

The more the supplier is integrated in the early phases of the project, the more the supplier's selection process moves away from the one used in a client–supplier relationship, such as the traditional call for a new product that has already been technically defined [MID 97]. However, the potential supplier's level of expertise influences the timing of its involvement, as its ability to contribute should complement the level of internal knowledge (or knowledge in the already mobilized ecosystem) of the innovation project [HAN 99]. Thus, the more the project requires the input of the supplier to its knowledge base, the later it will be integrated, and vice versa [HAN 99, p. 77]. The greater the supplier's technological capacity, the less it will be challenged on it.

Before innovation projects, when the level of knowledge is still low and the design is not fixed, i.e. during the time of exploration and/or open creativity, the selection of a supplier is primarily made on the basis of a verification that integrates the experiences of previous collaborations [CRO 01] and strategic choices defined over the long term [DUS 95a, p. 147].

The earlier the integration takes place, the more necessary it is to assess the potential alignment of the two companies' strategies for the innovation project, in addition to the supplier's technical capacity.

Conversely, in the production phases – during the integration of suppliers once the innovation has been launched on the market, is in production or is close to being launched – the criteria for assessing suppliers are based on "*traditional*" criteria, i.e. the traditional quality–cost–delivery criteria [CAL 13, p. 33].

In the intermediate phase, the selection of suppliers is based more on potential and global professional competency, and then on global functional objectives [MID 97].

Relational alignment is important in the selection of suppliers because, in situations of uncertainty, the supplier's cooperation capacities can complement its technical capacities [MEL 13]. It is one of the factors that contribute to the success of the collaboration [HÅK 93]. This is particularly the case in client–supplier relationships that are already active, where familiarity and trust from past interactions help to limit relationship risks by serving as social control mechanisms [DEK 04, DEK 08, GUL 95, GUL 98].

Nevertheless, the academic literature does not agree on the impact of relational alignment with regard to how companies should select their suppliers: while some stress the importance of previous collaborations (ibid.), others suggest that they should be sought outside the existing network in order to renew innovation potential [BEN 18, MEL 13].

Although the criteria for the technical/technological assessment of the supplier have been widely studied[7], this is not the case for strategic and relational assessment [LED 11] – this requires the client to rely on an internal capacity in order to assess their own "soft skills", to identify their future needs and to formalize their own strategy before they can assess the alignment of the supplier [BID 99]. For the latter, a necessary first step must be strategically and operationally prepared before choosing a supplier [BID 99, p. 103].

Lastly, the selection process is in itself a learning process between a client and a supplier, due to the interactions that precede the integration decision [DEK 08].

This learning process first involves the assessment and constitution of the three alignments described above – it can lead to the formation of an integration mode that is specific to the client–supplier relationship of the targeted innovation project. It can also induce common learning about the technologies/products/processes of both companies [DEK 08].

The contribution of the project teams on the client and supplier sides (same potential) from this Open Innovation sourcing phase contributes to the construction of the object at the center of the cooperation, the specifications thus play the role of a mediating object between the collaborators [LED 14].

7 Without, however, linking them to the financial capacity of the supplier, in the case where the supply of goods/services resulting from the joint project requires a significant "ramp-up" (development of production lines, internationalization, massive recruitment, significant supply, etc., with cash gaps). However, nowadays, this assessment of firm capacity is an important question mark for large companies that do not have answers.

1.1.2. *Configuring the relationship for the innovation project*

A key phase between the sourcing of a supplier for an Open Innovation project and its official involvement, the relationship configuration phase of the innovation project is the subject of most studies regarding the governance of supplier involvement in innovation projects [DEK 08, SÄF 14].

According to the authors of such studies, the main activities that make a collaboration successful are those that take place during this configuration phase of the relationship [DOZ 96]. The aim is to build the initial conditions for cooperation on the project, both in terms of contractual framework and inter-firm coordination conditions.

1.1.2.1. *Which contract should be established?*

The existence and content of a contractual framework to carry out an Open Innovation project offers the possibility of strengthening (or weakening) the innovation potential that was imagined during the sourcing phase. Indeed, it makes it possible to specify the incentives on intellectual property – the potential financial and/or industrial exchanges for both the supplier and the client. In addition, the protection provided to everyone by the laws that apply to collaboration also influences their potential for openness and transparency.

To do this, the different forms of contractualization extend over a continuum that ranges from a transactional type of contract (engineering or subcontracting purchase model) to a "partnership" type of contract. In the first case, the objective of the contract is to be accurately performed or, in the event of failure, to serve as a basis for a dispute settlement [MID 96, p. 21]. In the second case, the subject matter of the contract includes the possibility of contingencies or changes after the initial agreement is signed; the contractual clauses encourage the parties to "escalate" the detection of problems as soon as possible (ibid.) and leave room for the creation of new common opportunities.

After defining the purpose of the cooperation, the second most important contractual element is the definition of the level of responsibility given to the supplier in an innovation project or its degree of autonomy. The latter feeds most of the typologies on the subject [LED 14]. The part that is carried out by the supplier can be divided into six levels of responsibility [HAN 99, MON 00, SER 17]:

1) *catalogue purchase*: the client acquires a product or service previously designed by the supplier who produces it; the client does not participate in the design or production;

2) *pure outsourcing*: the client is in charge of the entire design of the targeted innovation and the supplier is responsible for its implementation; it can have a certain flexibility on the choice of the production process;

3) *white box*: the supplier is consulted to give its opinion on the design choices made by the client who designs the innovation. It is responsible for the implementation of the targeted innovation;

4) *gray box*: the design of the product delegated to the supplier is jointly carried out between the client and the supplier; the supplier participates significantly in the design and production of part of the targeted innovation;

5) *black box*: specifications (often functional) are provided by the client; and the supplier provides expertise in order to clarify needs and then ensure the design, industrialization and production of an entire part of the innovation or the complete innovation;

6) *open box*: responsibilities are largely defined according to each other's areas of expertise; clients and suppliers are unable to define *ex ante* the objectives to be achieved, or even always the means to be used for the design and industrialization of innovation.

The adoption of one level of responsibility or another depends on the technological resources of the client and the supplier, with regard to the problems to be solved in the innovation project [BID 98]. The less the client masters the technologies necessary for the targeted innovation, and the more the supplier masters them, the bleaker the setup will be: the client must rely on the expertise provided by the supplier in order to carry out their innovation[8]. A good knowledge of the level of responsibility left to the supplier, in terms of development, is a prerequisite for defining an appropriate management method for the relationship [BEN 04].

---

8 The other interest of the delegation of responsibility is the transfer of risks to the supplier – a subject that is studied in the literature on supply chain management and industrial networks. Moreover, being in a black box does not presuppose that this "box" is inaccessible to the client.

A third key element of the form of contractualization adopted is the distribution of the fruits of collaboration on the innovation project. The study on alliances and vertical relationships focuses on the issues of "pie sharing" and its equity [JAP 01].

Research on Open Innovation is generally less common on the subject – it holds little interest in contractual commitments that exploit the results of the innovation project. Instead, it is interested in the economic model of the Open Innovation project. It is considered that the economic model of the Open Innovation project must be in line with the economic model of each party [CHE 03, WES 14]. When it comes to contractual arrangements, this translates into a focus on intellectual property rights [LAW 11].

However, in research on the involvement of suppliers in their client's innovation projects, there is a bigger focus on the incentives put in place to reward the supplier's efforts. These incentives must ensure the cooperation of the supplier, while limiting both the client's innovation expenses and the risk of "lock-in" to the integrated supplier.

In addition, incentives through contractualization can range from direct remuneration for the supplier's efforts (transactional purchase and sale contract) to the promise of future gains (usually under conditions), in order to cover the investment made during the innovation phase [MAN 09, p. 128].

1.1.2.2. *How is inter-firm coordination organized?*

Good supplier relationship management is considered as a supplementary source of performance for the client firm [HÅK 06, LED 11]. It is reasonable to assume that the opposite is also true.

As part of the integration of a supplier into an innovation project, the management of inter-firm coordination begins with a phase of choosing the relationship management method for the project. It is then followed by its operational management as part of the project management [NPV 08].

The choice of the relationship management method consists of defining rules and methods that will allow a good coordination of exchanges between the client and the supplier on the Open Innovation project.

This choice is all the more important as the project is innovative. Indeed, when it proves impossible to define *ex ante* the objectives to be achieved, or sometimes even the means of implementation, "*the relationship is committed to the agreement of a general target that is necessarily inaccurate: on this basis, it establishes strong mutual dependencies, a common destiny; it then asks that everyone does not focus on the detailed respect of an initial exhaustive contract, but to jointly accomplish a task that incorporates internal hazards and external disruptions*" [MID 93, p. 178].

This choice is made according to the level of responsibility, the level of uncertainty of the project (in relation to the technical difficulties to be solved and its maturity) and the agreement, sometimes negotiated, between the client and the supplier of the project. This involves setting up mechanisms that can have two roles: a role in information and knowledge transfer between the two parties, and a role in risk control.

Operational management must make it possible to secure these two roles throughout the life of the project, by monitoring its maturation and evolution on the object of cooperation, as well as on the changes in personnel and functions involved within each firm. It then requires the use of a large number of devices with the most varied profiles.

Research on this subject has noted that the organization of meetings and the use, or even planning, of information exchange alone are not enough to ensure the success of the innovation project [SÄF 14]. Others also noted the particular role of space in coordination methods and knowledge transfer, particularly that of direct physical exchanges through project floors [GAR 96, LAK 06a], collaborative innovation spaces [FAB 17] and physical objects [BRE 13, GAR 96, KOS 09, LED 14].

Nevertheless, the mechanisms for exchanging information between clients and suppliers involved in an innovation project require a careful balance so that their impact remains positive on the performance of the relationship [SOB 02]. The manner and context of these exchanges also impact the nature and effect of the knowledge exchanged: it can either be solely transferred from one body to another or generate joint learning through the quality and extent of the inter-firm relationship [ROS 14].

In addition, risk management systems are pivotal to managing the relationships surrounding an innovation project, even though they mainly originate from an intra-organizational level. They generally involve the measures that have been put in place to limit the risk of dissemination of knowledge that has been exchanged and created, and are initially based on a contractual basis. Then, there are assessment mechanisms related to the project deliverables: product design and manufacturability, availability of parts, respect for delivery, quality and costs, etc. [VAN 01, p. 183].

Lastly, methods for assessing the performance of the supplier itself can be set up. They can then take the form of joint teams, workshops or a relationship assessment tool. When implemented through shared tools accompanied by socialization mechanisms between those being assessed and those assessing, they are considered relevant by both parties [LED 11], as well as a source for performance [COU 07]:

– for the client, because they make it possible to identify critical issues in the short term and correct them as early as possible in the innovation process; in the medium term, they make it possible to continuously improve the supplier's performance; in the long term, they make it possible to improve its supplier base for future projects;

– for the supplier, because the assessment makes it possible to clearly identify the performance criteria expected by the client and thus improve its organization to move from a standard supplier status to that of a key supplier providing innovations. This assessment also allows it to lower its coordination costs when joint subjects are highly complex.

1.1.3. *Internally managing the supplier's involvement in the project*

While it is widely stressed how closely inter-firm coordination methods are linked to internal management methods within each firm [BID 99, LAK 01, LAK 06b, MAN 09, TAK 01, VAN 08], the question of the impact of internal client–supplier regulations on the success of client–supplier Open Innovation is a subject that has been little explored.

This is an important subject from the get-go of any collaboration, requiring significant efforts from both companies, but more particularly from the client firm [TAK 01]. Thus, according to Van Echtelt *et al.* [VAN 08], the main difficulty in projects involving clients and suppliers is the lack of managerial competency in the management of this type of relationship. Too often, the client focuses on supplier selection, and defining their role and time of involvement [CHE 11, p. 82]. There is then a difficulty in knowing how to generate and maintain their own capacity to manage their innovation cooperation with their suppliers, and conversely for a supplier with its clients.

Many studies highlight the importance of involving the highest levels of management for the success of innovation projects, particularly when they are carried out between a client and a supplier [ROY 04]. However, this link is not enough either. The different interactions between the staff from both parties on a day-to-day basis are also important. Involvement must take place throughout the chain of command, from the highest levels to the most operational levels [BEN 07, LAK 06a, p. 816].

Moreover, when the impact of internal integration methods on the performance of the relationship is studied, the results show that the lack of consistency of internal methods is often negative [MAN 09, p. 437], even though it may turn out to be positive [MER 13]. The issues related to the intra-firm consistency of an innovation project, particularly when it comes to the R&D–production interface, may sometimes not be transposed to the relationship; this is a field that remains to be investigated [ENG 14, SÄF 14].

Lastly, from a long-term perspective – beyond the limits of a specific innovation project – the ability of a client firm to internally manage its relationships with its innovative suppliers is considered a key resource: its relational competency [DYE 98] in innovation.

Van Echtelt *et al.* [VAN 08] proposed that the long-term management of the client–supplier interface consists of three activities: (1) motivating them to strengthen or maintain specific knowledge or develop certain products; (2) exploiting their technological capabilities; and (3) assessing their development performance. For a client firm, this means knowing how to manage each supplier separately and appropriately [DYE 02]. This ideally allows access to the status of a "*preferred client*", which opens access to innovation resources that are specific to the relationship [SCH 12].

It is also a question of making the link between the strategic space of the client firm (policy of opening up its innovation and management of the panel of suppliers to be involved in innovation) and the operational space relating to the innovation projects in cooperation [MAN 09, VAN 08] (Figure 1.2).

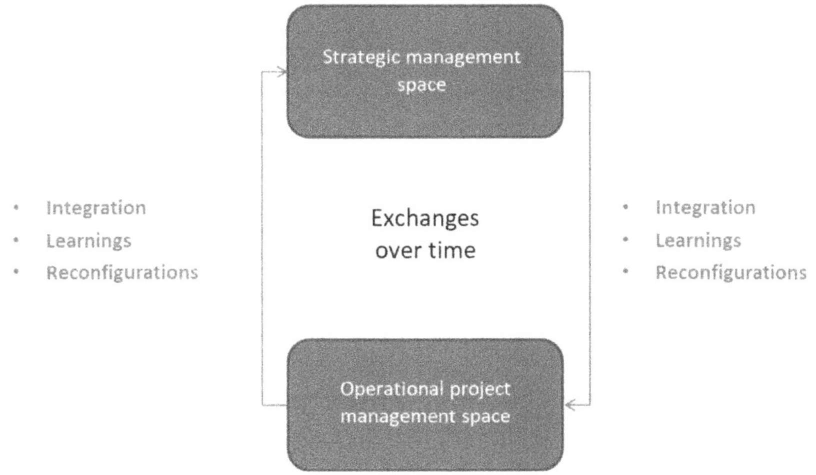

Figure 1.2. *Integration of strategic and operational activities in the client–supplier Open Innovation relationship*

1.1.4. *The pivotal role of Purchasing in Open Innovation*

At the heart of these interactions, the literature particularly focuses on the role of "*innovation intermediaries*", both internal and external to the firms involved in Open Innovation. Among the first are the "Open Innovation Champion" [CHE 06a], the firm's stakeholders who contribute to the implementation of Open Innovation approaches [MOR 11].

The work on these intermediaries suggests that they help organizations to meet and understand each other, and then facilitate knowledge transfers between organizations involved in an Open Innovation project [CAV 12, GAR 09, MON 17, ROB 12].

Among these intermediaries, the Purchasing function is beginning to play an increasingly studied role [BEN 18, MAY 17, MIK 18, PAT 17, SER 18b], and one that is increasingly recognized by managers.

Today, the involvement of the Purchasing function in the sourcing of Open Innovation, as well as its participation in innovation projects, is considered as a "best practice" [SCH 10]. Moreover, the more a firm opens up its innovation, the more essential its role is with regard to reducing uncertainties in the upstream phases of projects [CAL 00, p. 37].

Nevertheless, its positive impact is still being discussed [JOH 09, MEL 14] and seems to be as much a matter of its integration with other entities involved in innovation [BRA 13] as of the performance monitoring system of the Purchasing body involved in the innovation project [ATU 95].

The conventional view of this function is based on its central role in supply chain management, namely to ensure that the goods/services provided by suppliers are delivered with the right level of quality, at an acceptable price and within the agreed deadlines. It also has the role of seeking new suppliers, assessing and selecting them, maintaining contacts with them and negotiating the terms of the agreements agreed upon with them. Thus, traditionally, the buyer is both a negotiator and an organizer of the supply chain, who intervenes once the product is perfectly designed [CAL 00].

With the growing call for suppliers to participate in their clients' innovation projects, and thus the involvement of this function, a less significant study (by size) than that on the involvement of suppliers in Open Innovation has developed around the EPI: "early purchaser involvement". These studies involve describing the new and different roles that Purchasing entities can take in the development of their firm's innovation, whether these roles are internal or external.

Although this role in the success of innovation cooperation is generally considered quite small in relation to the role of R&D entities [ATU 95, BRA 13], these studies seek to describe it and identify its impact that is integrated with other internal entities as well as with the supplier firm.

For EPI researchers, Purchasing entities are recognized as being the most relevant for managing the selection process of a supplier to be integrated into an innovation project [JOH 09, PET 05, VAN 08] – particularly on the commercial side, which makes it possible to establish trusting relationships and ensure commitment between the parties.

Indeed, Purchasing is considered to be the most appropriate body to assess the availability and relevance of external resources for integration into the innovation process [WYN 99].

Their knowledge of the competitive market allows them to recognize the value of the knowledge found there and to facilitate the firm's access to it, along with its acquisition [TRE 98]. Their involvement in this process has two objectives of cost reduction: "*action on development costs, by taking advantage of the economies of scale and scope that can be generated by the early involvement of specialist suppliers (in innovation projects), and action on industrialization costs by taking manufacturability problems into account from the design phases*" [CAL 00, p. 35].

Purchasing is also considered, along with R&D, as the most relevant internal function to manage suppliers involved in innovation projects [DOW 92, MAY 17, MEL 14, VAN 08]. Moreover, beyond the selection process and in connection with this role of interface with suppliers, their role extends to the management and coordination of internal interfaces [BEN 04, JOH 09, MEL 14, SCH 10, SER 18b, TAK 01], and more particularly, their integration with R&D entities [BRA 13, CLA 91, WYN 03].

Furthermore, the Purchasing function not only contributes to innovation through its action on cost, quality and delivery issues, but also by linking innovation and external resource management strategies [MEL 14, SER 18b]. Thus, in addition to having an operational role, it also has a strategic role and establishes a link between these two levels of innovation cooperation management [VAN 08].

Some studies have also observed that there may be a distinction within Purchasing departments between entities that are in charge of life-cycle (or strategic) sourcing and advanced (or forward) sourcing [JOH 11, SCH 10]. The first is in charge of strategic sourcing with a stronger commercial and supply chain approach towards quality–cost–delivery, and a stronger link with internal customers than the second which focuses on new product development projects (ibid.). The latter is responsible for supplier selection activities for a given innovation project, selection of the necessary technologies and participation in the project itself [MEL 14, WYN 03][9].

9 These two "advanced buyer" roles can also be divided into two distinct entities, depending on whether they are involved upstream of the project or as part of the innovation project [JOH 11, SER 18b].

Its role is therefore that of the facilitator, who completes the project manager and plays a pivotal role in coordinating between internal and external systems [BEN 04]. It acts as a "troubleshooter" in cooperative innovation projects, both at the operational and strategic level [MEL 14]. Nevertheless, these different roles that an Advanced Purchasing body can play in the management of vertical innovation cooperations are still not much studied and need further work [LUZ 15, MEL 14].

1.2. The dynamics of client–supplier cooperation in innovation

The dynamics of an inter-firm relationship can be defined as the characteristics of change (and its "opposite", continuity) over time, and the factors that generate these characteristics [CRO 10, p. 636].

This subject is discussed in studies on strategic alliances [DAS 02, DOZ 96] and client–supplier relationships [DWY 87, LEE 12]. Such studies make the link between the temporal aspect of the relationship and the change/stability dichotomy in its governance and atmosphere – the atmosphere being the ambiance specific to the client–supplier relationship, which is part of a broader strategic context: its environment [HÅK 82].

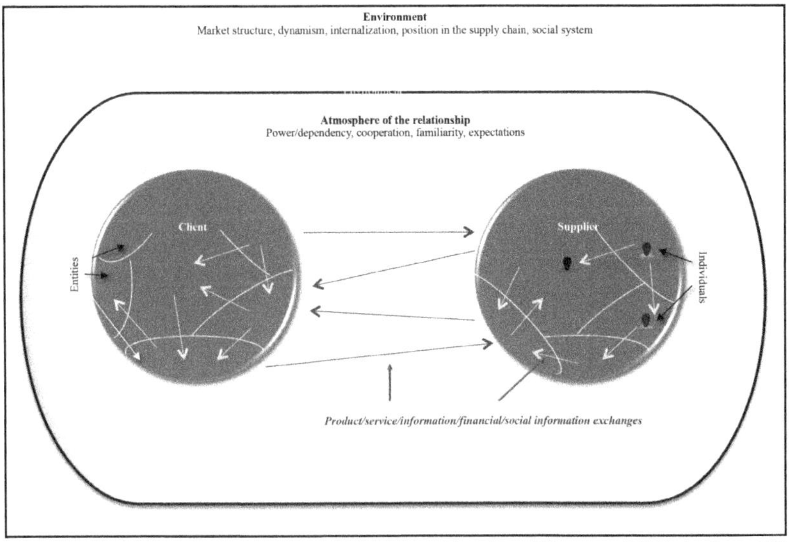

Adapted from Håkansson and IMP Project Group, 1982

Figure 1.3. *Atmosphere and environment in the IMP group's interaction model (adapted from [HÅK 82])*

This link is revealed as being particularly critical. For example, Reuer and Ariño [REU 07] noted that strategic alliances can undergo a wide variety of changes during their lifetime. These changes may lead to the modification of the terms of the contract (21% of the alliances studied), the composition of the alliance's management or organizational committee (18%), the introduction of mechanisms to monitor collaboration (15%) or the expansion into new areas of collaboration (15%). Overall, 44% of the alliances they studied have experienced at least one of these types of changes.

In the study of inter-firm relations, the impact of cooperative experience on the performance of the relationship is widely demonstrated [ZOL 02, p. 707]. Each firm's ability to make adjustments both in line with the dynamics of the relationship, and external to the relationship, in addition to maintaining their consistency throughout the relationship, is considered a cause of success (or conversely, failure) of cooperation, in the case of client–supplier cooperation [GUL 05] and strategic alliances [DOZ 96].

Similarly, in the case of development projects involving suppliers, both the subject matter of the cooperation and the contractual structure may be subject to changes during its implementation [GAR 94, MAN 09, SEG 03].

Furthermore, in order for an Open Innovation project between a client and a supplier to be efficient, an important evolution of the internal workings of both can be equally required [MID 97].

However, the dynamics of the inter-firm relationship is the subject of a limited amount of research [CRO 10, p. 642] and only few are interested in it in the context of innovative situations [YLI 14]. This is one of the major limitations of the work on Open Innovation between the client and the supplier, and therefore a development path for future research [LAK 14, SEG 03, YLI 14].

But for this to happen, it is necessary to take an interest in the issue of time.

Time can be understood both as the experience acquired and as the evolution of the environment, challenges and resources that are external and internal to companies, without these evolutions necessarily being the consequences of the relationship (even though they have a direct impact on it). Time is therefore registered beyond a purely Newtonian sense, that of the clock.

In order to understand the dynamics of client–supplier Open Innovation, in the following, we have explored these different aspects of time in the inter-firm relationship, as encountered in the main theoretical approaches that have been used for the research that we present in this book [SER 15].

Thus, after having discussed the effects of the experience of the dyadic relationship on client–supplier innovation cooperation, we will focus on the two schools of dynamics of inter-firm cooperation. We will then briefly discuss the issues related to the dynamics of the environment of the Open Innovation relationship: changes in the context and in the firm strategies.

1.2.1. *The effects of the experiment on the client–supplier relationship*

The repetition of contacts and projects within a client–supplier relationship can lead to the development of organizational routines that are specific to the client–supplier relationship. These routines can be considered as coordination methods that appear over time.

They mainly involve the establishment of informal mechanisms for managing the couple, which can replace the formal mechanisms that were initially defined, or fill a gap.

For example, the constant routine of information exchange episodes within an inter-firm relationship leads, after a while, to a clarification of each other's expectations and the distribution of roles to be played within the client–supplier relationship [HÅK 82].

Personal relationships also replace inter-organizational relationships according to the formal roles of the entities involved; psychological contracts replace formal contractual arrangements [RIN 94]. When these routines help to adjust the methods of cooperation management in order to lead it into a virtuous spiral of successful cooperation [DOZ 96, p. 74], they are an important source of performance for the client and the supplier: through the quality of their cooperation, they derive more benefits from the relationship than they expected to find.

In addition, a second mechanism related to the iteration of exchanges and collaborations between organizations is the learning effects that reinforce familiarity between companies. First of all, they have an impact on the value created by cooperation [GRA 96] and thus allow the continuity of the relationship, all while participating in the congruence of organizations' perceptions [BAR 07].

The experience acquired between the two parties makes it possible to reduce information asymmetries between them [WIL 85], both regarding the behavior of one towards the other, and the knowledge they have of one another [VAN 09]. It thus makes it possible to strengthen the innovation capacity of each firm by increasing the knowledge base made accessible over time.

Learning about the relationship with the other (compared to learning about bases of knowledge, innovation) leads each party to a more precise perception of the risks of opportunity by every party; and when this perception is reduced, it leads to an increase in trust between them [RIN 94]. These learnings also facilitate the communication process between the two parties [SOB 02].

The more organizations exchange with each other, for all possible types of exchange (information, goods, services, money, social) that take place in accordance with what was expected (or in accordance with their *ex post* interest), the more willing they are to let the other act independently, without fear of losing control [RIN 92].

Van de Vrande, Vanhaverbeke and Duysters [VAN 09] thus showed that companies with common experience adapt their contractual governance so that there are more commitments in the sharing provisions.

The need for control systems also decreases if the experience has been positive, for example, the legal form of the contract makes less use of capital-intensive systems [GUL 95, GUL 98, VAN 09]. As confidence increases, there is a decreasing need for complex contractual arrangements [RIN 94], control arrangements [GUL 95, RIN 92] or formal information sharing arrangements [REU 07]. In these cases, these are supplemented or replaced by organizational routines [ZOL 02].

However, routines and experience can have a crippling impact on the couple's agility. There may come a time when the expectations of one and/or the other are so institutionalized that they are no longer questioned by either [HÅK 82]. These routines can then lead to a loss of relationship performance due to increased coordination costs [GUL 98] or even a lack of rationality regarding decisions [HÅK 82], or a decrease in creativity [AHU 01, SAP 04].

A relationship that is too close can limit the possibilities of "thinking outside the box" and simply create new combinations between everyone's knowledge [NOO 07]. However, Deeds and Rothaermel [DEE 03] also showed that there is a U-shaped curvilinear relationship between the age of R&D alliances and their performance.

1.2.2. *The evolution of inter-firm relationships*

When an inter-firm relationship is analyzed within the framework of a common project (an alliance as an innovation project), the reference episode to identify its formation begins with the identification of one party by another and lasts until the contractualization of an agreement, including its negotiation.

The influence of what has happened between the two companies before this formation phase is theoretically considered in the study of the relationship. However, pre-formation interactions are not integrated into the analysis of the relationship *per se*. The role of the supplier as a "silent designer" during the sourcing episode is not considered. Only its effect is considered as a context element.

Yet, the dynamics of a relationship around a common project can begin well before this first phase of meeting the expectations of both parties, even though there have not been any episodes of client–supplier interactions. A relationship between two companies begins before their agreement is contractualized.

Interactions between the client and the supplier on an Open Innovation project can start as early as the pre-contractual awareness phase and first appointments [DWY 87], from the formal contract at one extreme to the coffee meeting between the staff of each entity at the other extreme [ROY 04].

The relationship is based on an awareness of the possible synergy of resources of one party identified by another, in order to strengthen its own resources. Episodes of information exchange between parties (visits, exploration of opportunities to work together, etc.) can precede episodes of exchange of products and services for a considerable amount of time [HÅK 82, p. 25].

The beginning of the relationship is achieved through the first interactions, which consist of an exchange of information between them. It could even be extended to the first contact between an influential member of one of the companies with the other firm, reputation being an integral part of the assessment criteria for one firm by another [ZAH 98]. These initial interactions make it possible first to verify the congruence of the expectations of each party. If the relationship survives these very first interactions, the dynamics are then set in motion.

In the study of inter-firm relations, two academic schools consider the dynamics of the couple. They are particularly interested in the evolution of the atmosphere of the relationship, and more particularly in trust and the power/dependency relationship. One takes a "life cycle" approach to the interest, and the second takes a "teleological" approach [POO 04, p. 24, THI 99, VAN 95, p. 130].

1.2.2.1. *The "life cycle" approach to the relationship*

The first school proposes a sequential approach to couple dynamics (Figure 1.4).

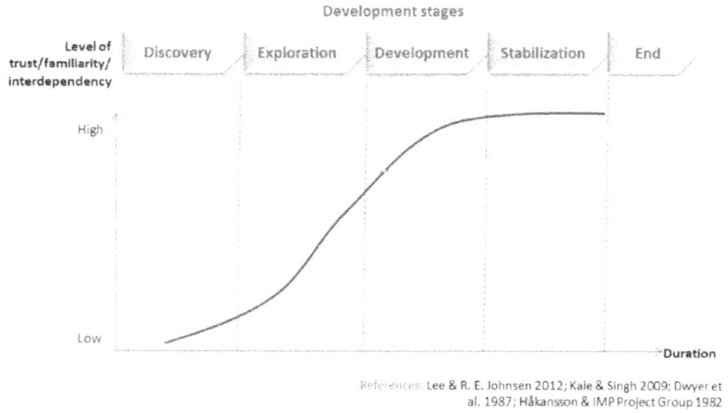

Figure 1.4. *Life cycle approach to the inter-firm relationship (sources: [LEE 12, KAL 09, DWY 87, HÅK 82])*

It defines the life cycle of an inter-firm relationship as the succession of different phases of maturation of that relationship. For Dwyer, Schurp and Oh [DWY 87], these are the phases of awareness, exploration, expansion and engagement. For Kale and Singh [KAL 09], these are the phases of cooperation formation (including partner selection), definition of the governance and structuring of the alliance, post-formation management of the alliance and the end of the alliance. For the IMP group [HÅK 82, p. 300], these are the pre-relational, early, developmental, long-term and final phases.

The transition from one phase to another is made through the positive influence of the experience of cooperation on the characteristics of the atmosphere [DWY 87, HÅK 82, p. 292, KAL 09, LEE 12]. The atmosphere of the relationship is gradually changing [HÅK 82], and for some companies, the time required to move from one phase to another can take increase or decrease [CAN 99, p. 456].

By summarizing this work, an inter-firm relationship develops in four main successive stages:

1) *discovery* represents the stage where there is an awareness and first assessment of the potential synergy that leads to the choice of one another [DWY 87, KAL 09]. There is a "unilateral consideration of potential exchange partners" [DWY 87, p. 15], but no commitment between the two parties [HÅK 82, p. 300];

2) *exploration* is the process of initiating discussions, the first negotiation and the first interaction phase of the relationship [DWY 87, KAL 09, LEE 12]. It represents the stage where the first exchanges make it possible to better assess the resources of the other, by comparison to its own resources;

3) *development* is the state characterized by the multiplication of interactions, the development of a specific capacity to work together and the reduction in uncertainties that are related to the relationship, in particular by the development of investments that are specific to the relationship [DWY 87, HÅK 82, LEE 12];

4) *stabilization* is the state of the balance of power within the relationship. This state allows for a long-term mutual commitment based on established and mutual trust [DWY 87, p. 19, LEE 12]. Information exchange mechanisms are then given priority over control mechanisms [GUL 95, KAL 09, REU 07, RIN 92].

This sequential perspective has the advantage of facilitating the understanding of the development of a relationship. It also corresponds to the development of best practices: that of inter-firm "partnerships", where both parties commit themselves to a sustainable and win-win relationship[10]. In these cases, mutual trust increases both reciprocal commitments [MOR 94] and the performance of the relationship [DON 07], which in turn helps to increase trust.

1.2.2.2. *The teleological approach to the relationship*

The second school proposes a teleological perspective of the evolution of the couple's relationship; teleology being the philosophical doctrine according to which the purpose or aim is the ultimate cause to guide the movement of an entity [VAN 95, pp. 515–516]. According to this perspective, an organization, which is the relationship here, evolves with a final objective in mind. As a result, and since its inception, the inter-firm relationship evolves through a continuous cycle (Figure 1.5) of achievements, assessments and adjustments [DOZ 96, RIN 92].

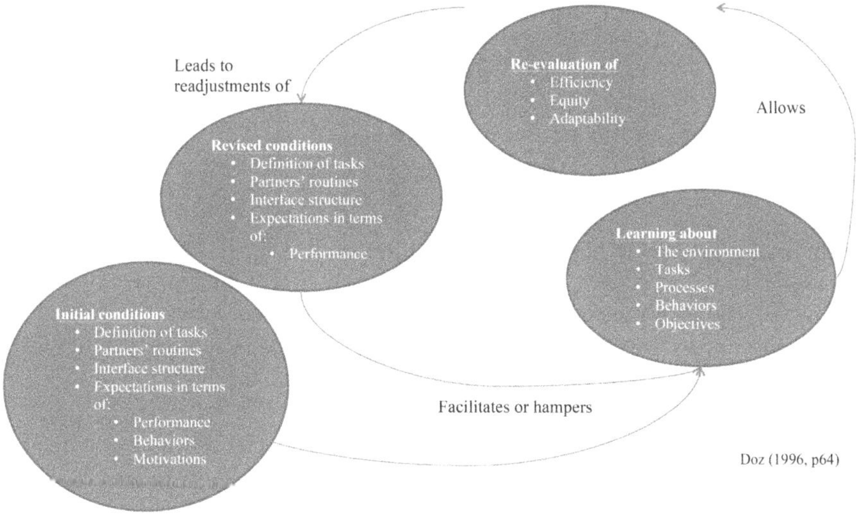

Figure 1.5. *Teleological evolution of the inter-firm relationship*

---

10 In this same vein, Traitler and Saguy [TRA 09] proposed that the proper management of an Open Innovation project in "partnership" mode can be represented by the succession of four successive stages, from the initial contact to the creation of joint value: (1) an essential dialogue ➔ (2) the establishment of an atmosphere of trust ➔ (3) the construction of goodwill ➔ (4) value creation.

Each episode executed allows each of the firms, through their interactions with each other, to develop their knowledge of one another and their ability to work together [DOZ 96, SLU 11]. Each episode of assessment is an opportunity to gage the performance of the collaboration, future perspectives and their dependence on each other.

Each adjustment episode corresponds to the implementation of measures to respond to this evaluation. An organizational configuration specific to the relationship is initially adopted, based on everyone's expectations and resources. The initial conditions and the very first cycles of development of a relationship are important in its history.

Through the commitments made by both parties, followed by the verification of the implementation of these commitments, learning is encouraged or hindered [DOZ 96]. It is through these commitments that both interpersonal and inter-organizational trust can thrive [ZAH 98].

Thereupon, the conditions of the cooperation, and its methods of organization are adapted to each cycle. It is also during these cycles that the development of the relationship is achieved through increased collaboration; a decrease in collaboration is also possible and might lead to a failure of the cooperation [DOZ 96, p. 79, DOZ 75, RIN 92].

As a result of these experiences and successive learning, each relationship follows own trajectory [SEG 03, p. 187]. The ability of each firm to make adjustments to its operations, in addition to maintaining its coherence throughout the relationship, is considered a cause of success or failure of the cooperation [DOZ 96, p. 74].

The two schools coincide in the sense that they are predictive theories that describe an end point and identify initial conditions [DOZ 99]. For both of them, the development of the relationship does not continue to grow (when it grows) indefinitely. For both schools, it can achieve a balanced state of contributions through a mutual sense of equity [DOZ 96], and power through increased knowledge and recognition of each other and their importance to one another. They also agree on the possibility of the relationship discontinuing, in other words, its termination regardless of its state of development.

These two schools of dyadic dynamics also coincide with the association between the atmosphere and governance of the inter-firm relationship. In addition to an increasing level of familiarity, each phase or cycle of evolution is characterized by a specific level of trust and interdependence between the companies. These three characteristics evolve over time; they influence and are influenced by the governance of the client–supplier couple.

1.2.3. *The evolution of the relationship environment over time*

During the relationship and outside of the joint innovation project *stricto sensu*, events can also impact the functioning of the relationship by making it take a new turn through changing the situation in a sustainable or temporary way. These events occur in what we can define as the environment of client–supplier cooperation in innovation.

The cooperation environment encompasses both the external environment, such as markets, competitors, governments, and the internal environment, which is the strategic context for the partnership within each organization [DOZ 96, p. 69]. In general, companies have a fairly similar perception of the external environment (ibid.).

Over time, their respective strategic contexts may evolve in a more or less compatible manner, as well as converging or diverging. As a result of changes in environments and people, as well as new resources resulting from cooperation or the emergence of technological breakthroughs, everyone's strategies and organizations may vary [AND 84]. In these cases, it is a question of adapting their operating methods and coordinating their adjustment [GUL 05], if cooperation continues to apply to their own strategic context.

In addition, over time, conflicts, communication issues or changes in each other's expectations are inevitable [INR 94]. These differences eventually affect the terms of the relationship. During the resulting renegotiations, new contractual agreements may be established, correcting all or part of pre-existing contracts.

If this negotiation leads to finding converging interests, the development of the relationship continues (ibid.) and can improve performance. As such, conflict can have a positive role in partnerships; when it is resolved it becomes a source of information, mutual learning and leads to alignment of cooperation objectives and governance methods [AND 84].

Lastly, in the context of innovation projects, such as supply chain life cycle management, the maturation of the project (the transition from one phase to another) often leads to the involvement and disinvolvement of different functions in each firm. For example, in a purchase–sale project, the teams in charge of supplier selection (sourcing) may differ from the teams in charge of the supply, one being involved before the other. In an innovation project, Research laboratories are involved upstream and then pass information on to Development teams, which will then pass it on to the Methods and Operations teams that are downstream.

These changes in functions involved in innovation cooperation are another source of potential discrepancies that affect the atmosphere of the relationship, owing to internal discrepancies between the objectives and interests of these entities [DOZ 87, p. 18, JOH 08, p. 282, WHE 92]. The most common example is the transfer of relays in the client firm, between an R&D department and a Purchasing department, in the case of an innovation project tackling pre-industrial phases. This can be particularly destructive when there has been no Purchasing engagement upstream.

1.2.4. *In conclusion, interpretation is needed for reading the dynamics of Open Innovation*

Contemplating the dynamics of client–supplier Open Innovation, including their links with relationship performance, raises multiple challenges that have been identified and individually considered by the academic literature.

These challenges still need to be addressed from a relational and dynamic perspective. In light of the academic literature, there remains a need for interpretation that will allow this complexity to be understood. This interpretation should make it possible to consider the determining factors and levers of the performance of such relationships over time. It will then have to consider the trajectory of the relationship, as well as the innovation situation. This is the purpose of the next chapter.

2

Conditions and Impacts of Client–Supplier Open Innovation Governance

How is client–supplier Open Innovation organized? Under what conditions is it organized? And what is the impact on the performance of the relationship? All of these questions were what guided the context of interpretation for the governance of client–supplier Open Innovation.

In order to create this basis of interpretation, having not been identified in the literature, we began by comparing different academic theories from the perspective of the different players of this collaborative innovation (different positions in companies, client companies, as well as suppliers, of different sizes, of different sectors of activity, in addition to institutional and intermediary players).

From this first research work, we have derived a first model and hypotheses on the links between its different elements.

This model was then tested with a sample of 160 supplier companies cooperating with a client on an innovation project (Figure 2.1). They were asked to describe in detail how this cooperation was managed, under what conditions and toward what aims.

This allowed us to detail our model in a very efficient way, in addition to confirming or refuting the importance of certain elements and the hypotheses linking them.

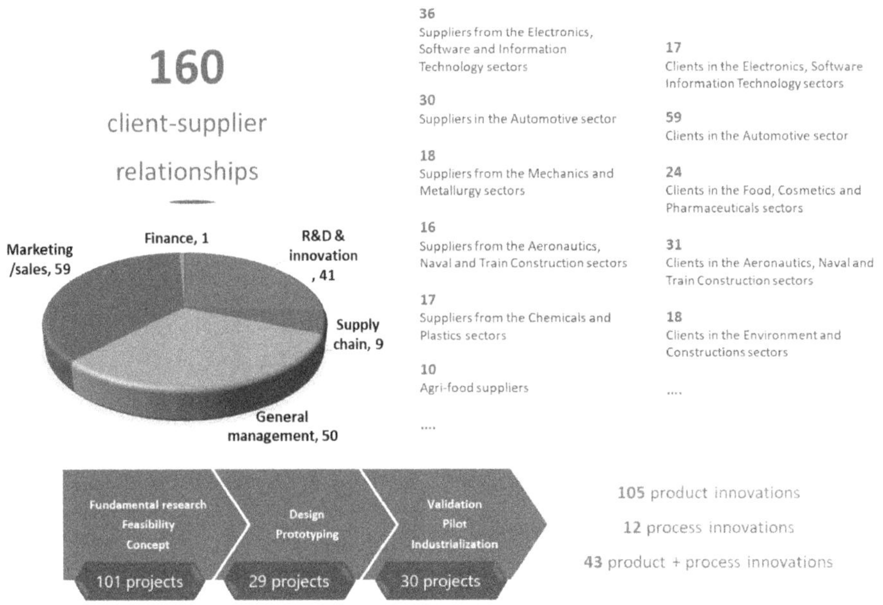

Supplier companies

Client companies	TPE	PME	ETI	Large group	total
TPE	4	1			**5**
PME	3	12	3	2	**20**
ETI	4	8	8	3	**23**
Large group	19	45	39	32	**135**
total	**30**	**66**	**50**	**37**	**183**

Distribution of client-supplier pairs according to their size

Figure 2.1. *Summary description of the sample of the quantitative approach*

In this chapter, we describe this final model element by element – presenting both its components from a theoretical and an operational perspective. We begin by describing the performance of the client–supplier Open Innovation relationship. Then, we present the elements and configurations of its governance. Next, we present the atmosphere of the relationship and its links with governance and performance, before indicating what influence the innovation project may have on this same governance.

This chapter concludes by proposing which forms of governance would be best adapted to the different conditions encountered in client–supplier Open Innovation.

2.1. A benchmark: the performance of the relationship

2.1.1. *What is performance?*

The performance of an Open Innovation cooperation can be measured in different ways: through its longevity, the number of patents filed, the R&D expenses incurred, the speed at which new products are marketed, the impacts on the balance sheet or income statement of both parties, the achievement of the objectives of individual companies, etc. [AHU 01, DAS 00].

We are interested in the *relationship* performance, in other words, the degree of achievement of the objectives of both the client and supplier who cooperate in innovation; the client–supplier relationship is then considered as an entity in its own right. It is this coupled performance that we examine.

The very reason for inter-firm cooperation is to generate a quasi-return: it is through this cooperation between these firms and specific investments that productivity increases, in addition to reaching a level of resource that it would not have reached without the other [BAN 11].

Furthermore, this cooperation can generate an additional source of profit that is jointly generated by the two companies [DYE 98], one that goes beyond this quasi-return. This source of additional profit, the relational return, is a resource that is specific to each client–supplier partnership that succeeds in generating it.

We also propose to measure the performance of a client–supplier cooperation in Open Innovation through these three aspects, which correspond to:

1) *the effectiveness* of the relationship, in other words, the achievement of the formal objectives of the cooperation. The objectives of both parties relate to both the joint innovation project(s) and to the exchanges related to the

client–supplier relationship[1]. They cover a wide spectrum of objectives such as: meeting costs, quality and delays for innovation projects; achieving the objectives of the innovation project or program; developing everyone's ability for innovation through joint learning; the strategic value of the relationship, etc.;

2) *the efficiency* of the relationship refers to the ability to make the optimal use of shared resources. It corresponds to the application of the ability to work together in order to achieve the objectives of the dyadic relationship. It can be assessed through the transaction costs[2] due to face-to-face contact between the two parties, and/or through the satisfaction of individuals working together;

3) *the proactivity* of each firm involved in the duo corresponds to its level of commitment in the relationship, in other words, its belief that the other party is so important that it must make every effort to maintain their relationship indefinitely [MOR 94]. This is reflected by an ability to anticipate the expectations of the other and to implement the resources that allow it to propose solutions without an explicit request [LED 11] or even invoiced. It is a way of approaching the specific relational return of the client–supplier couple that are involved in an Open Innovation project.

This performance that we use as a benchmark is not the performance of the supplier for the client (or the client for the supplier), nor that of its involvement in the innovation project, nor that of the joint innovation project. This is indeed the performance of the *client–supplier relationship* in Open Innovation.

1 The level of trust in an inter-firm relationship is not a relevant indicator for assessing the performance of vertical innovation cooperation. If trust allows cooperation to take place and encourages its performance, rather it is an indicator of the state of development of the relationship. Thus, it is normal for a level of trust to be low at the beginning of a relationship – without the partnership being ineffective – relative to the level measured after a successful innovation project or a successful business cycle between two companies; moreover, trust is also not of the same nature at these two moments. The performance indicator related to trust would instead be the change in trust over a reference period. Confusion may arise in particular from the notion of a "goodwill trust", in other words, the willingness of the other party to do more than is formally expected [SAK 92], which is considered here, when translated into action, as proactivity.

2 Transaction costs are all costs relating to the entire project that make it possible to achieve the desired objective; they include both the costs of assessing the means to be implemented and those related to efforts to reduce the uncertainties associated with the project and the nature of the investments specifically made to achieve it.

Measuring the performance of the relationship

Obtaining objective data on the performance of an inter-firm relationship is considered all the more complicated as this data is often considered confidential [GUL 05]. Since subjective assessment is strongly correlated with objective assessment in studies of inter-firm alliances [KAL 02], this measurement can be made on the basis of a qualitative assessment carried out by managers involved in these relationships [KAL 07, SLU 11, ZOL 02].

Based on the three aspects of performance – effectiveness, efficiency and proactivity – we use measurement scales from the literature on the performance of alliances [KAL 07], client–supplier exchanges [BOY 95] and innovation projects involving suppliers [LED 11].

Effectiveness is measured by considering the extent to which the main objectives of the relationship are met [KAL 07], at one time just as at the end of the cooperation.

Efficiency can be seen through the ease of cooperation and good team work of the two companies [BOY 95], and also in negative terms by identifying the presence of glitches in interactions, in other words, the "lack of sharing of information and knowledge" [HOO 99].

Proactivity is approached by measuring the level of commitment of each firm in the partnership. It is then a question of looking at whether one firm spontaneously offers ideas or resources to the other without solicitation, whether it offers new business opportunities or whether it carries out its tasks towards the other with the greatest enthusiasm [LED 11].

2.1.2. *Three levels of performance measured in client–supplier Open Innovation relationships in France*

Our quantitative study allowed us to identify three levels of relationship performance[3]:

3 The reliability of the measurements were validated through the calculation of Cronbach's alpha of the nine variables, initially measuring each of the performance components. This is 0.781 (average of the 45.6375 scale, standard deviation 7.17481 and variance 51.478), which indicates a good reliability of the scale [CAR 10, pp. 61–62].

– “*outperformance*” is marked by the highest average levels of each of the performance components. On a scale from 1 to 7, all variables are above 6; only the client’s proactivity drops to an average of 5.84;

– “*performance achieved*” involves 51% of the sample. Each variable drops by one step in relation to those of outperformance;

– “*low performance*” is characterized by the lowest levels of the average of each of the variables, as well as the highest variances. Only the average of the “efficiency” variable exceeds level 5 (“moderately agree”) when the client’s proactivity is measured at below average levels (level 3 corresponds to “moderately disagree” and level 4 to “neither agree nor disagree”).

Performance level

Performance classes	Average / Variance	Effectiveneness	Efficiency	Proactivity of the supplier	Proactivity of the client	Work-force
Class γ - outperformance	Average	6.29	6	6.22	5.84	34
	Variance	0.21	0.27	0.26	0.27	
Class α - performance	Average	5.59	5.62	5.30	4.60	82
	Variance	0.39	0.46	0.47	0.74	
Class β - low performance	Average	4.82	5.18	4.17	3.17	44
	Variance	1.59	1.22	0.92	1	

Scale from 1 to 7

Figure 2.2. *Distribution of performance components*

Additional analyses[4] have identified that among these variables, the client’s proactivity towards their supplier seems to be the first determining factor of performance.

The client’s proactivity is therefore strongly correlated with each of the other three aspects of performance. Thus, while it can be considered as a measure of the performance of the relationship as a result of the commitment of either firm in the couple, it also reflects the impact of this commitment of the client on the success of the cooperation.

4 A principal component analysis of the performance components revealed that the variable measuring the client’s proactivity explains 51.71% of the variance. This was confirmed by an analysis of the correlations between these same variables.

Lastly, these analyses of performance components also revealed the strength of the link between the effectiveness and efficiency of the relationship. The achievement of the established objectives of cooperation is very highly correlated with the quality of inter-firm coordination.

2.2. Governance of the relationship, elements, forms and influence on performance

2.2.1. *What elements constitute the governance of client–supplier Open Innovation and how do they influence performance?*

The governance of a client–supplier Open Innovation relationship consists of all the formal and informal contractual and relational mechanisms which ensure and regulate the various exchanges within the client–supplier relationship that are related to the Open Innovation project.

These mechanisms are heterogeneous, even disparate. They include processes, tasks, tools, routines, contracts, dialogue modalities, etc. They refer to both inter-organizational and intra-organizational mechanisms [TAK 01].

Together, they form the governance of each client–supplier partnership interacting on an Open Innovation project.

Among this heterogeneity of systems, we can distinguish three main families that make up this governance: (1) contracts (formal and legal definitions of inter-firm mandates); (2) management mechanisms (use of formal and informal mechanisms to transmit information and control mandates); and (3) human resources (or the players of the duo, functions and individuals). We describe them in the following sections.

2.2.1.1. *The contracts*

Contractual governance limits the legal scope of the relationship through a formal framework, where cooperating firms mutually agree on their expectations, rights and obligations [KAL 09, MAC 80, REU 07, SOB 02]. The contracts must cover the different situations that may arise [OUC 80] during the different episodes of the relationship. Even though the parties do not necessarily follow the terms of the formal contracts, they do provide them with common “rules of the game”.

Contractual governance is made up of the various contracts and legal arrangements put in place between a client and a supplier. Contracts are put in place in order to protect the relationship against possible opportunistic behavior from both the client and the supplier. They may not be used in a relationship or for a specific project – the absence of a contract is then considered as the least complex contractual governance method.

A contract is specified by (a) the choice of the legal form of the agreement, (b) the period of validity of the agreement, (c) the scope of the agreement and the implications for each party and (d) the degree of standardization or personalization of the agreement [SOB 02].

The scope and implications of the agreement are embodied in contractual clauses that can be found in all legal forms of innovation cooperation [REU 07], from the purchase and sale contract to the creation of a joint venture through the various forms of partnership.

The contractualization of cooperation between a client and a supplier on an innovation project can take several forms. To discuss these, we have selected variables that touch upon the legal form of governance of the innovation project described, based on the classification of Kale and Singh [KAL 09].

These clauses can be distinguished according to (1) whether they contribute to safeguarding the relationship or (2) whether they define the sharing of resources and results of cooperation. In the following sections, we present these two types of clauses and identify their influence on the performance of the relationship (Figure 2.3).

	Condition of influence on the performance of the relationship	
	Negative influence	**Positive influence**
Safeguard clauses	• Absence • Too many constraints	• Presence
Sharing clauses	• Imbalance	• Equitable sharing • Voluntary imbalance • Commitment

Figure 2.3. *Influences of contractual arrangements on the performance of the relationship*

2.2.1.1.1. Contractual safeguard clauses

The safeguard clauses included in a contract make it possible to limit the uncertainties associated with opportunistic behavior, by giving each party the ability to impose its will on the other without its consent [MAC 80, WIL 75]. Examples include clauses setting out the rules for resolving conflicts or for settling disputes in the event of non-performance of the contract by a party, clauses limiting the use of information exchanged and produced or clauses allowing one to audit the other.

The risk of sanctions has a positive effect on the relationship, both by forcing cooperating firms to remain focused on their common objectives and by building trust based on deterrence within the relationship [GUL 95].

The choice of a legal form of governance using a capital-intensive mechanism (acquisition of an interest or creation of a joint venture) can be compared to a contractual safeguard mode: the legal inter-firm link created by these forms is considered stronger than that created by other forms (purchase contract, license, co-development contract, etc.). The risk is therefore not that of the sanction, but of the cost generated by the unraveling of the link, the exit from the contract, which may be higher than the cost of a sanction agreed upon in the contract.

The presence of safeguard clauses in the contractual governance of cooperation entails obligations for each party and the possibility of exercising unilateral power over the other [MAC 80].

Example of contractual safeguard clauses

The following contents can be specified in the clauses of an inter-firm cooperation contract [PAR 93, REU 07] in order to set up means of coordination (the first three) or to have the other honor its commitments (the following four):

– The regular written report of all relevant exchanges and transactions between the two companies.

– The right to access and audit all relevant accounts and records of the other firm.

– Prior written notification in the event of termination of the agreement by either party.

– The use of arbitration in the event of a conflict.

– The possibility of legal proceedings in the event of failure to fulfill contractual obligations.

– The possibility of late penalties.

– The definition of the end of the agreement.

The constraint each party imposes on the other impacts the performance of the relationship, by ensuring its continuity to the extent that this constraint is limited. As soon as they are perceived by one or the other as being too constraining or coercive, such as limiting their room for maneuver, the impact on performance becomes negative [HAU 10].

Indeed, as soon as the safeguard clauses instability in the power relationship, the risks of opportunistic behavior from the more dependent firm multiply [PRO 89], using part of its resources to seek to regain leeway.

In addition, the safeguard clauses have a curvilinear impact on the performance of the relationship [HAU 10]: a minimum is required to ensure the continuity of the relationship, but above a certain threshold, the multiplication of these clauses reduces the performance of the relationship.

2.2.1.1.2. Contractual sharing clauses

The sharing clauses define the contributions of each firm, their way of contribution and the rules for sharing the results of cooperation between companies: method of remuneration, sharing of intellectual property, follow-up rights and obligations, etc.

By setting the rules for pooling their resources and the distribution of roles and responsibilities, the objective is both to agree on the scope of cooperation and to limit the risks of opportunism, with the cooperation strategy being superior to the exit strategy.

In the case where the subject of the cooperation is clearly defined, such as in the case of a contract for the procurement of off-the-shelf products/ services, the content of these clauses is the subject of a "traditional" negotiation which involves identified exchanges of products/services and financial flows depending on the quality, costs and delivery times of the exchanged products/services.

In the case of Open Innovation projects, the purpose of the cooperation is rarely clearly defined. Sometimes, it cannot even be defined. The commitment of these clauses then forms a substantial stumbling block with regard to the formation of the contract – especially when it comes to the sharing of intellectual property.

Generic rules for sharing results within a client–supplier Open Innovation project can be:

– a firm's proportional contribution determines its share of the results;

– each firm receives half of the results;

– the supplier is paid a fixed fee regardless of the result;

– the supplier is paid a fixed fee if the project is successful;

– the supplier is remunerated on future sales, based on a predetermined margin percentage;

– the supplier is remunerated by a fixed fee in the case of success and on future sales.

As to the question about intellectual property, there are four types of generic rules:

– new patents (and others) go exclusively to the supplier;

– new patents (and others) go exclusively to the client;

– the client and supplier apply joint patents;

– the client and supplier share new patents and licenses according to each other's expertise/industry.

The most common principle for establishing these sharing clauses is equity rather than equality [JAP 01] – equity corresponding to a sharing of results that is proportional to the sharing of pooled resources [BEN 04, RIN 92]. A fair balance between resources involved and profit sharing is sufficient and equality is not necessary for sharing to be considered fair [BLA 64].

This principle of equity contributes to the quality of the relationship, both through its impact on satisfaction with the collaboration and the perception of justice that encourages continued collaboration [JAP 01, RIN 94].

In addition, the rules for sharing results may be deliberately unbalanced by one player to its disadvantage, in order to ensure the commitment of the other. This imbalance can be a selling point to differentiate oneself from the competition when it offers interesting alternatives.

A medium-sized European laboratory may thus convince biotechnology companies to carry out joint innovation projects by leaving them all intellectual property rights in American territories. To differentiate itself from the major competing laboratories, it thus offers its potential partners a sharing of results in which it abandons part of its prerogatives.

Beyond the economic and rational aspect of sharing, it can also form a "sociological debt" that makes it possible to ensure the commitment of the other in return [RIN 94].

Lastly, in contracts overseeing an innovation project, result sharing clauses may bind the parties beyond the safeguard clauses. This involves including clauses anticipating resale rights. These consist of committing to the order/supply of the innovation resulting from the project and the means to be implemented by one and/or the other party in order to ensure this supply. The difficulty then lies in setting a price or a method of calculating the price that is sufficiently incentivized for each of the parties.

This commitment, beyond the sharing of the intellectual property generated by the project, is an explicit promise to continue the relationship beyond the time of the project, in addition to its continuity [DWY 87, p. 19]. It can be a source of additional performance for the relationship, if it does not lead to an imbalance in the relationship, either by creating unsustainable obligations (too high/too low a price, impossible-to-meet deadlines, etc.) or by increasing "hold-up" risks (locking oneself in a monopolistic situation, abuse of a dominant position, etc.).

2.2.1.2. *Management mechanisms*

The systems that organize and manage a relationship between companies are heterogeneous by nature [DUM 09]. The mechanisms for managing a two-party relationship represent the use made of various formal and informal methods that govern interactions within the duo. These interactions occur at the inter-organizational, interpersonal and intra-organizational levels. They relate to the various exchanges that take place there: financial flows, goods, services, information, etc.

The central issue of these mechanisms is the coordination of the relationship, in other words, the search for coherence in the work accomplished by all the individuals and functions involved in the Open Innovation project.

In order to assess the impact of relationship management mechanisms on relationship performance, we propose classifying them according to whether they participate in (1) information and knowledge exchange, and/or (2) relationship control.

Some methods can be used for one type of mechanism or another. For example, the planning of a project makes it possible both to exchange information on the methods of its implementation and to monitor it by referring to it. Management mechanisms are active at both organizational and individual levels.

In the following sections, we present these two families of mechanisms and identify their influence on the performance of the relationship (Figure 2.4).

	Condition of influence on the performance of the relationship	
	Negative influence	**Positive influence**
Information exchange mechanisms	• Absence • Too big of a multiplication	• Face-to-face • Intermediary objects • Articulation (learning) • Lack of articulation (project success)
Control mechanisms	• Absence of negative incentives • Strong negative incentives • Bad assessment • No assessment of the relational rent	• Positive incentives • Adjustment of management mechanisms

Figure 2.4. *Influences of management mechanisms on the performance of the relationship*

2.2.1.2.1. Information exchange mechanisms

To address information exchange mechanisms, we assimilate knowledge exchange with information exchange just as we assimilate a contributor, to varying but indistinguishable degrees, to building the innovation and cooperation capacities specific to each client–supplier relationship [HAT 99].

These mechanisms make it possible to transfer and assimilate information. By making them clear-cut, they promote the development of learning both within and between organizations [NON 95]. These learnings involve both the understanding of the functioning of the other party, and of one another, as well as the knowledge related to innovation, whether it comes from a transfer or joint learning [ROS 14].

By relying on the tools and processes that enable the management of the relationship, its steering and its implementation, they ensure a process of standardization and sharing through the dissemination of information. These include, for example, the implementation of both formal and informal meetings, the publication of reports, computerized data exchanges (letters or specific data exchange platforms such as integrated management software packages, product life cycle management tools or computer-aided design software), the use of intermediary objects (models, prototypes, diagrams, etc.) and also quality initiatives or co-location of personnel on project floors or in collaborative innovation spaces [FAB 17, LÔ 16]. They are implemented both within and between organizations.

These information exchange mechanisms also ensure a process of standardization (through the tools and processes that allow the management of the relationship, its steering and its implementation) and sharing (through the dissemination of information such as formal or informal meetings, or the publication of reports or memos).

They are implemented both within and between organizations. These different mechanisms have a positive effect on learning within the relationship when they are well articulated, or alternatively, when they come from the success of joint projects [SOB 02]. Moreover, through this potential contribution to one of the two major objectives of an innovation cooperation (the development of learning that is used for the project or will be used in another context or project), they can have a positive impact on the performance of the relationship [SLU 11].

The mechanisms of increasing direct exchanges also contribute to improving the performance of the relationship. This has been very widely demonstrated, such as in studies in the 1990s that focused on client–supplier relationships and sought to identify the differences in performance of these relationships in the Japanese automotive industry when compared to other countries, whether at intra-organizational or inter-organizational levels [CLA 91, DYE 96, NON 03, TAK 01].

For example, Dyer [DYE 96] showed that the increase in face-to-face exchanges reduces the time to market of new models. Although for Barnard [BAR 38], this should be at the management level, Doz [DOZ 87] pointed out that the operational level is also important for the success of innovation cooperation.

On the contrary, the excessive multiplication of these mechanisms can also negatively influence the performance of the relationship. By bringing together knowledge bases and relationship norms, they limit the possibilities of new combinations, with individuals' perceptions becoming redundant [AHU 01, HAM 12, NOO 07, SAP 04].

Similarly, the frequency and intensity of communications have a curvilinear (inverted U-shaped) impact on the budget of innovation projects between client and supplier, and also on the resulting product costs [HOE 05]. In addition, these information exchange mechanisms must be carefully balanced so that their impact remains positive on the performance of the relationship [SOB 02].

2.2.1.2.2. Control mechanisms

Relationship control mechanisms represent the non-contractual means implemented to ensure that the interests of each party are safeguarded [GRA 96], in other words, the specific objectives of the cooperation and the individual objectives of the cooperating companies.

Their role is to ensure that exchanges comply with, on one hand, contractual mechanisms, and on the other hand, with the strategies and standards of each entity involved in the relationship. They are mainly based on the formal risk management systems of companies at the level of cooperation. Das and Teng defined these mechanisms as the regulatory processes by which the pursuit of the mutually compatible interests of both parties is made more predictable [DAS 98, p. 493].

The mechanisms used are, for example, the processes for monitoring and managing innovation projects, Customer Relationship Management/Supplier Relationship Management, charters, budgeting and assessment methods deployed in organizations, and also the influence methods (incentives and deterrents) implemented.

The verification of the compliance of one party with the other's mutual commitments is carried out through the continuous assessment of the relationship by each party and the possible implementation of adjustments [DOZ 96, LED 11] and influence strategies, in order to ensure compliance with the objectives [HAU 10].

Through the definition of assessment standards and the development of knowledge of the relationship's environment, the needs of one's firm, one's entity and the capacities of the other to deliver according to its own expectations, the importance of the cooperating entity is estimated by one another [GRA 96], as well as the levels of risk in the relationship. These mechanisms reflect the value system of the firm, of the entity or even of the person who establishes and implements it.

These assessments determine the choices made by the partner firms [MAR 93]. For example, the lack of valuation of the relational return can lead to the multiplication of opportunistic behaviors or the termination of relationships when alternative solutions present equivalent capacities; such valuation can be done both through formal mechanisms such as the assessment of sunk cost or, to a lesser extent, TCO (total cost of ownership), or informal or even cultural mechanisms such as considering the importance of the relationship[5] in the performance of a business relationship, or the fear of loss of reputation in the event of opportunistic behavior.

5 This mechanism is "naturally" present because of the business cultures of certain regions of the world: this is the case in Confucian tradition – the importance of *guanxi* in business relations in China is illustrative; it is the same for German *Standort* culture.

The issue that then arises does not just regard the presence of assessment mechanisms, but also of their relevance and *a fortiori* of their ability to have a positive or negative impact on the performance of the relationship [BER 83].

Assessment mechanisms are the backbone of control mechanisms through their role as gatekeepers for the safeguard strategies to be implemented [INR 92]. They make it possible to define the level of cooperation, the positive or negative incentive strategies to be implemented, and also guide the adjustments to be made in inter-organizational as well as intra-organizational mechanisms.

They are realized at the beginning of the relationship by communicating the procedures and operating rules from one (often the client) to the other, then at the beginning of the project by defining the roles and responsibilities of each firm [PER 13]. Subsequently, throughout the Open Innovation project, both of them regularly and structurally monitor the progress and performance of the relationship with the other, and sometimes with one another (through steering committees, satisfaction surveys, etc.) [BRE 13, SLU 11].

The difficulty then comes from the way in which these are mechanism(s) operated. An assessment mechanism that is not aligned with the objectives of cooperation, or the absence of assessment, can lead to a distorted evaluation of the conformity of the exchanges with these objectives and therefore also lead to inappropriate adjustments.

In addition, the formal control mechanisms attached to assessments are considered as drivers of the performance of inter-firm cooperation [DAS 98, DEK 04, DOZ 96] – when their implementation costs are not too high, they are adapted to the context and used appropriately [BER 83, BRA 13].

Consequently, mechanisms for assessing the performance of the supplier's innovation capacity directly contribute to the performance of the relationship, when assessment is accompanied by socialization mechanisms between those assessing and those being assessed, such as joint teams and joint workshops [COU 07]. The impact of a single control mechanism may depend on the information exchange mechanisms that can complement it.

In addition, incentive mechanisms can either be formal, through the implementation of contractual arrangements or management tools, or informal, through influence strategies. The latter particularly affect trust and commitment within a relationship, and thus influence cooperation and compliance between partners in the same channel [HAU 10]. They can be practiced in a positive or negative way.

Example of informal incentives

Informal incentives are reflected in the attitude of individuals towards each other in order to influence their engagement in the project. In the examples below [HAU 10], these incentives can be positive (the first two examples) as well as negative (the following ones) when the client's personnel:

– Focus on the positive impact of successful innovation cooperation on the supplier's business.

– Highlight what they will bring to us in return for the supplier's participation.

– Do not fail to refer to the contractual commitments.

– Make biased interpretations of agreements between the two companies.

– Are very clear on the penalties that will be imposed on the supplier if it does not succeed in its role.

– Threaten to disrupt the business relationship if the supplier is unable to fulfill its role.

Box 2.1. *Informal incentives*

Control mechanisms through positive incentives can be practiced through the review of contractual governance by strengthening commitments [MAN 09] or by simply increasing encouragement towards people. These positive incentive mechanisms will contribute to strengthening the effectiveness of the relationship through the generation of higher engagement generated [HAU 10]. However they can also reduce its efficiency if their implementation adds complexity or imbalances to the relationship.

Control mechanisms through negative incentives can be practiced through finicky interpretations, biased interpretations of contractual governance, or the use of threats against the other organization or its personnel [HAU 10]. It appears that such mechanisms generate positive results when operating at low levels, even though they are generally counterproductive and discourage both cooperation and compliance with the objectives pursued [HAU 10].

Lastly, a final category of control mechanism consists of implementing adjustments in the management systems used within the two firms. The aim is to adapt them to the situation of cooperation. This adaptation must be done between the two firms, and must also occur within the firms themselves. The arrangements of the parties to an innovation cooperation must be adapted to the situation of the cooperation and be able to manage the integration of the results generated on a tailor-made basis [MAN 09].

2.2.1.3. *The human resources involved*

Human resources are made up of the players of the client–supplier relationship that are involved in the Open Innovation project. These players, which are functions of the firm and/or individuals, are both carriers and converters of knowledge, users of management mechanisms and implementers of social and organizational integration of the relationship, as well as of the innovation project.

In an Open Innovation project, there are essentially several functions (and arguably, individuals) of each firm that are involved in the relationship. Each function has a role in its firm and the individuals who belong to these entities carry its roles and related expectations [GAB 87].

When they are in charge of the different stages of innovation, from the exploration of new ideas to their exploitation in a supply chain, these functions and individuals are often separated both in terms of organization and location.

Sell–purchase activities as well as production–delivery activities are rarely carried out by the same entities, let alone the same individuals, that are in charge of activities related to the innovation project or program. This separation between supply chain and innovation-related activities is found both in the corporate world and in the academic world, and both recognize the importance of interfacing these entities for the performance of innovation and relationships.

In addition to this separation, each of its functions, and also each individual, can have its/their own objectives and interests [DOZ 87, p. 18, JOH 08, p. 282, WHE 92], and can thus also have governance mechanisms and evaluation standards that are specific to it/them. The time of their involvement in the different phases of innovation corresponds to the importance of the different logics that predominate within organizations: technological logic for R&D, economic logic for Purchasing, market logic for Marketing, logistics logic for Supply Chain, quality, production, etc.

Doz has thus noted that technological partnerships were decided by head management and carried out by middle management and technical specialists; the lack of liaison on these subjects between these different entities then created significant risks of misunderstandings and frustrations within the relationship, which could lead to the end of the partnership [DOZ 87][6].

This implication, while it affects the present time of the project, also impacts its future when this implication coincides with a selection/choice episode. The consultation of a given entity, its capacity to influence and its decision-making power are all elements that highlight the object of cooperation – beyond the issues relating to its innovation and production capacities, or to the culture of firms or entities. The mobilization of one function or another can have different impacts, depending also on the phase of the project. Ring and van de Ven [RIN 1994] have that negotiations tend to proceed better when lawyers are absent. Moreover, in Chapter 1, we showed how the involvement of the purchasing function could, under certain conditions, contribute to the success of Open Innovation projects when it is involved upstream.

2.2.2. *The governance of Open Innovation: a complex blend*

We have just seen how the governance of an Open Innovation relationship is composed of an extremely varied set of mechanisms. Each

6 These issues at the individual level are part of a level of analysis that we do not address here, despite the recognized importance of individuals in facilitating or limiting innovation and collaboration capacities in the context of client–supplier innovation cooperation [CHE 06b, HÅK 82, TEE 07]. The study's conclusions on this subject are also contradictory, as the instability of a team during a new product development process can cause losses of information and knowledge [HOO 99], and can also, conversely, renew the creative capacity of teams caught in routines [DAY 10].

has the ability to encourage or discourage cooperation within a relationship. However, this influence is neither linearly or individually positive nor negative, and even less so on a collective level [HUI 11].

Describing each of these mechanisms allows us to retrospectively understand what happened. And yet, the analysis of each individually taken mechanism may present contradictory results [DUS 95b, p. 508], particularly when it comes to understanding its influence on performance.

In order to fully understand the governance of Open Innovation, and prepare for action, it is necessary to consider each mechanism in relation to the others.

The governance of the Open Innovation relationship is a complex system where each mechanism must be considered in relation to the others. It is not possible to know its parts without understanding the whole.

It is also necessary to analyze this complex system as a whole, the intra-firm in relation to the inter-firm, the formal in relation to the informal, etc.

We propose to consider the governance of Open Innovation as a whole. As such, it may have distinct configurations, the plurality of mechanisms implemented forming a configuration that is specific to it [HEI 03]. In addition, this governance influences the performance of the relationship. And each configuration can be linked to different levels of performance.

Through our quantitative survey, we validated these assertions and identified four governance configurations for client–supplier Open Innovation. From the simplest to the most complex configuration, there is:

– *the "Unstructured" configuration*: this is an example of governance reduced to its simplest form: a contract with few clauses specific to the supervision of an innovation project, or an absence of a contract, combined with a virtual absence of formal control mechanisms, and also of information sharing between the two firms;

– *the "R&D project" configuration*: includes contracts of intermediate complexity with a project horizon. It does not have bureaucratic or coercive control mechanisms. It is basically the only configuration where only R&D functions of the two companies are involved – and with a very high frequency;

– *the "Exclusive partnership" configuration*: this is highly complex in all its components: the management mechanisms are the most numerous and are a combination of inter-firm and intra-firm. This governance is partnership-based because it provides for mutual commitments beyond the project, if it succeeds, and in these cases, they are exclusive. Many functions are involved;

– *the "Developed partnership" configuration*: this is the configuration with the most complex contracts, with the largest number of clauses safeguarding the relationship and mutual commitments beyond the project. It is the configuration in which the client makes the greatest use of control mechanisms, both through bureaucracy and coercion. The supplier, on the contrary, uses third parties to support it in managing the relationship. Many functions are involved.

This survey also statistically confirmed the link between governance and performance identified by other management researchers [NPV 14].

The analysis of the governance–performance distribution reveals that the last three types of governance have 42% to 50% of "performance achieved" and are also represented in "low performance" and "outperformance" (between 25% and 29%). On the contrary, the "Unstructured" type of governance appears to be the least conducive to "outperformance".

In order to better understand what causes these differences, we detail the components of each of the four types of governance as follows.

2.2.2.1. *A – "Unstructured" governance*

This type of governance is the most represented in our sample (60 out of 160 partnerships studied). It is more marked than other configurations by the weakness of its structure, both at the contractual level and at the management mechanism level.

Here, the innovation project is mainly the subject of a purchase contract or is part of the "traditional" contracts in progress. There are few contractual clauses specific to the safeguarding of the project. Sharing of results is limited: when there is a commitment from one to the other in the event of success of the project, it is limited to a commitment from the supplier to develop its industrial capacity. There is also little production, and therefore little sharing, of intellectual property in these projects.

Relationship control is not carried out through common processes for managing the innovation project or relationship either. The client tends not to communicate its rules and procedures.

Informal attitudes of cncouragement are low. On the contrary, the supplier firm is characterized by the presence of a person in charge of coordinating with its client (in 55 cases).

Lastly, the frequency of involvement of the client's Purchasing Department, the supplier's Sales department and the R&D functions of both companies is high.

In this type of governance, the absence, or near-absence, of a contract dedicated to the innovation project is the biggest issue, while the importance of the involvement of the purchasing–sales and R&D functions shows that cooperation is very real. Is it due to a lack of recognition (or willingness to recognize), through an innovative-type of contract of the project (and/or the project itself), on the client's side? Or conversely, is it due to such a high level of trust between the two companies that it is not necessary to formalize mutual commitments?

The distribution of this type of governance with regard to the performance of the relationship (35% "low performance" and only 10% "outperformance" – as for the other three, they range from 25% to 29%) would encourage us to favor the first hypothesis with regard to the second, even though the second is possible for part of the sample[7].

This result extends the results of previous research that has shown that the performance of R&D contracts was better for detailed contracts in situations of low uncertainty, and conversely for more incomplete contracts [FRE 04, SAD 00]. Clauses that are too vague are demotivating [ING 03]. Through our survey, we show that in practice, it is possible to go even further: the absence of a contract does not make it possible to achieve outperformance [SER 18a].

7 We are weighing in here on the limitations of the large sample survey exercise. Although we can formulate and validate statistically consistent hypotheses, it is rarely possible to go as far as a high level of detail.

2.2.2.2. *B - "R&D project" governance*

This type of governance is characterized by the marked and bounded nature of the contractual and relational arrangements specific to collaborative innovation projects[8].

Although there is a contract for the innovation project, it is not a purchase contract but rather a consortium or co-development contract. It is in this class that there is the most public funding (19 cases out of 33). The contract system differs from the others in that it has the fewest resale rights. The sharing of results is done through the mere distribution of results, not on future sales. When there is intellectual property production (16 cases), it is shared through joint patents or cross-licensing agreements.

Although there is a sharing of defined roles and responsibilities and joint project reviews on the innovation project described, there is no particular relationship or project management process for each firm. Coordination is done through these joint reviews and through a person who coordinates the relationship on the supplier side.

Lastly, there is a very high relationship level with the R&D entities of each firm – the highest of all types of governance – and a relatively lower level of contact than in all others with Supply Chain management functions.

2.2.2.3. *C – "Exclusive partnership" governance*

"Exclusive partnerships" are characterized by the highest complexity of management mechanisms, in addition to significant contractual complexity.

The contracts that oversee the 29 innovation projects described tend to be negotiated and customized. They may be purchase or co-development contracts, but can also be *addenda* to existing contracts within the client–supplier relationship. The distribution of the results of the innovation project is mainly based on future sales. When there is intellectual property production, it only belongs to one of the two parties (in 13 cases out of 16, to the supplier). They contain resale rights from the client and supplier: the client is committed to buying exclusively from its supplier, and the supplier to develop its production capacity.

8 They seem to correspond to what is required in collaborative projects funded by the French and European public authorities.

With regard to relationship management mechanisms, the two companies follow a well-defined common process to manage their cooperation, at the level of the relationship and the innovation project. The management of each firm monitors the progress and performance of their relationship through regular and structured monitoring. This monitoring continues between the two companies through joint reviews.

The client and supplier each have a relationship coordinator (26 cases). The complete definition of roles on the project has been completed. The supplier also follows a well-defined internal process for setting up and managing its innovation cooperations, and its personnel share their experience and expertise in innovation cooperation with each other. Perhaps it is the supplier organization's ability to manage its Open Innovation that allows it to obtain exclusive commitments?

In addition, this governance is also characterized by client personnel who have the most encouraging attitudes and refer the most to their contractual arrangements. Perhaps this is a consequence of this exclusiveness? As long as you are linked to a supplier, you might as well make sure that you have the best interpersonal exchanges.

Lastly, there tends to be a strong involvement of Client Purchasing, Supplier Sales, R&D and Operations functions of both companies, as well as the General Management and Marketing of the Supplier. The Logistics functions of both firms are also involved.

The innovation targeted here is not only the production of something new but also the marketing of it, in a context of collaboration from design to production. This is reflected in the involvement of supply chain functions, in addition to innovation functions. And this is confirmed by the resale rights that have been contractually agreed.

Moreover, despite this multiplicity of stakeholders in each firm, interactions appear to be significantly more consistent than for other types of governance. It can be assumed that this is both necessary because of the strength of the contractual commitment and is possible because of the multiplicity of coordination mechanisms.

2.2.2.4. *D – "Developed partnership" governance*

This last type of governance (38 cases) is marked, more so than the others, by the complexity of contractual arrangements.

The legal form of the contract used on the innovation project largely takes the form of a customized, negotiated purchase-and-sell or co-development contract. They contain project-specific clauses: the end of the contract is scheduled and regular reports are required.

The main method of sharing results is mainly based on future results (increase in supplier sales). Thus, contracts tend to include clauses containing "resale rights": a commitment by the supplier to develop its industrial capacity to deliver the products or services developed as part of the innovation project, and even sometimes a commitment by the client to purchase them from its supplier if the project is successful. Compared to "Exclusive partnerships", the client's commitments are more focused on a minimal volume.

Moreover, it is in this type of governance that there is the most intellectual property production (24 cases) – this tends to be distributed according to the expertise of each party.

Relationship control is primarily carried out by the client through the communication of its operating rules and procedures, regular information on how the supplier meets its objectives, and monitoring the quality of the relationship. Relationship management can also be achieved through the adoption of a quality approach (29 cases).

This is not the case for the supplier who does not follow processes to set up and manage its innovation cooperations. However, it regularly calls on external experts (lawyers, patent specialists, etc.) to help it in this cooperation (22 cases). This may be related to the importance of intellectual property production in this type of governance.

In addition, client personnel use contract reminders to encourage supplier personnel to cooperate – and they also sometimes make biased interpretations of their agreements.

In this configuration, there is often a coordinator within each firm to centralize the relationship with the other (22 cases).

Lastly, this cluster is characterized by a high level of contact between the different functions of the two firms, with significant informal exchanges. The level of contact with Supplier Sales and Client Purchases is the most important compared to other classes. Contacts are high with the R&D, Operations, Logistics and Quality functions of both companies, and the supplier's General Management and Marketing.

As with other types of partnership governance, this multiplicity of functions is to be linked to the importance of the contractual commitment.

2.3. The atmosphere of the relationship, elements, forms and influence on governance and performance

Atmosphere is defined by the Cambridge dictionary as the "the character, feeling, or mood of a place or situation". Being a benchmark for measuring a satisfactory exchange relationship [HÅK 82, p. 21], the atmosphere can be described in terms of the level of power-dependency between the client and the supplier, the level of trust and proximity–distance between them, or their familiarity.

In the following, we describe each of these elements that make up the atmosphere of an inter-firm relationship and establish how each is related to the governance and performance of client–supplier Open Innovation. We will then present the four atmospheres encountered in Open Innovation projects, including their links with the different levels of performance. Finally, we will describe the atmospheres identified for each type of governance.

2.3.1. *What are the components of the atmosphere of an inter-firm relationship, and how are they related to performance and governance?*

2.3.1.1. *Trust*

Trust is at the heart of the study of the dynamics of relationships between organizations. It is considered an essential component of Open Innovation relationships [LIN 11, p. 13], one of the conditions for successful alliances [PAR 98]. It has a central, positive and evolving role in client–supplier relationships [DON 07]. Trust exists when a party is certain of the reliability and integrity of its trading partner [MOR 94].

Trust has been identified in different forms that we encounter here: trust relating to the knowledge of the other and the recognition of its managerial and technical skills [SAK 92], trust based on common standards, trust resulting from a calculation of (its) interests (positive or negative). It can be understood through both inter-organizational and interpersonal levels.

These two types of trust are built in their own right but are closely dependent, which explains why it is difficult to separate them even though they are not involved in the management of the relationship in the same way [ZAH 98]. In our work, we focus on inter-organizational trust[9].

By following Zaheer *et al.* [ZAH 98, p. 143], inter-firm trust is based on the expectations of one firm towards the other, namely:

– that it can count on the other to fulfill its obligations;

– that it will behave in a predictable manner;

– that it acts and negotiates honestly when it has the opportunity to behave in an opportunistic manner.

This definition stresses the possibility of betrayal by one of the two parties in the face of a higher prospect of gain [NOO 99a]. Trust limits the risks of opportunism without ever eliminating it. It is an "act of faith" in the sense that it is a matter of believing that the other is trustworthy, without ever being able to be assured of it [ZAH 98].

Trust is built through an interactive process based on reciprocity. It influences the behavior of stakeholders in their present actions and in their ways of considering and engaging the future [ARR 71]. By lowering the perception of uncertainties in the relationship [NOO 97], the establishment and development of trust leads to a reduction in conflicts and a facilitation of inter-firm exchanges.

9 Although interpersonal trust has the virtue of facilitating coordination between the staff of a client–supplies relationship and achieving results beyond what could never be achieved without it [MCA 95], it also presents the specific risk of collusion between the staff representatives of the two parties at the expense of their respective organizations: by privileging their interpersonal relationship to the interest of the inter-organizational relationship, they remain perfectly loyal to each other [FRE 04].

Trust thus makes it possible to increase mutual commitments [MOR 94], but also increase the exchange of confidential data between partners on markets and technologies [HÅK 82], all while limiting the risks of their disclosure. This trade facilitation promotes learning, particularly in innovative situations.

Trust also improves supply chain efficiency [ZAH 98, ZHA 11] and the effectiveness of the resources used, in particular by reducing exchange-related transaction costs [ZAH 98].

Thus, even though the relationship of trust and performance is still not obvious (the first one having a largely subjective aspect and the second one having an objective aspect), it is widely demonstrated in the academic literature [DON 05, 07, ZAH 98]. All quantitative studies on inter-firm relationships validate the positive influence of trust on partner performance [DON 07].

Moreover, trust can be considered as a complement to relational governance [DWY 87, GUL 95, HAU 10, MAC 80] or as a potential substitute for legal protection mechanisms [GUL 95, SOB 02]. Trust positively influences the development of relationship-based governance rather than protection, which translates into more cooperation, joint action, coordination, involvement of partners in accepting relational norms: information sharing, harmonious resolution of conflict, communication [DON 07].

Trust also makes it possible to encourage the implementation of incomplete contracts by replacing this incompleteness, as is the case in the context of innovation [FRE 04, p. 77].

Thus, on the one hand, the bigger the trust[10], the less there is a need for control mechanisms [GUL 95, RIN 92] such as the threat (and application) of coercive measures or the establishment of legal or organizational protection mechanisms.

10 It should also be noted that the influence of interpersonal trust is considered to be limited on governance changes [ZAH 98].

On the other hand, control mechanisms also allow a high level of trust – their absence can lead to a decrease in trust.

Lastly, if trust does not exclude the use of control mechanisms, too much control kills trust, especially from a certain level of relational proximity, because too much control can lead to suspicion [NOO 97].

Some governance mechanisms have the capacity to positively or negatively influence inter-firm trust [MEF 08, PUR 09]. This is not only true for contractual governance and management mechanisms, but also for human resources.

Indeed, the lack of consistency between the different entities of the same party generally negatively influences trust [DOZ 87]. However, sometimes, an apparent lack of consistency can be organized and have a positive influence. For example, the distribution between purchasing and R&D entities of specific roles to be played in innovation cooperation can promote the development of inter-firm trust: the former playing the role of the "bad cop" and the latter that of the "good cop" leads to the development of strong trust between R&D entities in both parties[11] [BRA 13].

2.3.1.2. *Asymmetry of interdependency or power imbalance in the relationship*

The notion of interdependency within a cooperation is based on the scarcity of alternative resources available to those that are accessible through the relationship [THI 59], broadly speaking.

The dependency of firm A on firm B corresponds to everyone's assessment of the importance of the relationship. The power of A over B is directly related to B's dependency on A [HÅK 82, p. 30]. And as the dependency of one firm on another increases, so does its commitment to the relationship [MOR 94]. The complexity of relational exchange lies in the fact that a large part of the power comes from this dependency of the parties on each other and that it grows throughout the development of the relationship itself [MAC 80].

11 All while preserving the control role assigned to the purchasing entity: the former can thus develop trust based on calculation when the latter develops trust that is based on "goodwill".

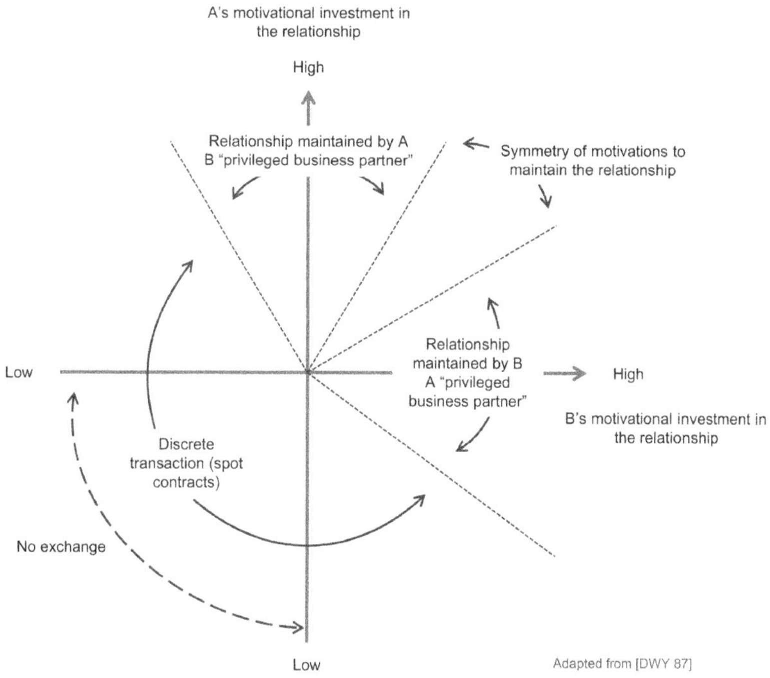

Figure 2.5. *Dependencies and interdependencies between client and supplier*

This dependency of one firm on the other results from the firm's assessment of the importance of the other partner in achieving its objectives, at the organizational level and not just the level of tasks carried out for cooperation [GRA 96] or the rarity of alternatives on the market [PRO 89].

Measuring dependency and interdependency

To assess the level of dependency between one firm and another, followed by the asymmetry, the principle is to look at how important the relationship is for everyone, and to what degree each perceives the opportunity to leave [AND 84].

In order to do this, the measures focus on the strategic nature of each other, the importance of the expenditures specifically made for the relationship, the difficulty of replacing the other and how big the cost of this possible change.

Box 2.2. *Dependency and inter-dependency*

The result of this assessment can be found, for example, in the mechanisms that are implemented in a differentiated way for "privileged business partners", who may be key account clients, strategic clients or suppliers, innovation suppliers, etc. (another result may also lead to a classification in "suppliers/clients at risk" – or sometimes partners considered at risk).

On the client firm side, this assessment and classification can be done according to the supplier panel management processes [KRA 83]. On the supplier side, this is generally done according to the importance of the client in current or future turnover – through the processes of managing the client portfolio.

Once the other firm is considered to belong to a specific category, it receives differential treatment [CAN 99]: contractual and relational arrangements are specifically implemented to ensure the performance of the relationship. Sometimes, it is the assessment process itself that is reserved for it [CAP 12].

In return, a firm treated in a differentiated way and receiving specific attention offers specific counterparties to the other firm. This is particularly the case when it is in a "first choice/preferred client" or "strategic supplier" category. Thus, a survey conducted by Bew showed that supplier companies offer their preferred clients "rare" products/services (in 75% of cases) and present them first with their innovations and new technologies (82%), in addition to offering them better prices (87%) [BEW 07].

On the contrary, partner firms considered as non-strategic are treated through standardized mechanisms that allow for a saving of resources in the processing of the relationship. The partner receives treatment in relation to its strategic nature.

While assessing the dependency of one on the other leads to the establishment of differentiated coordination mechanisms on either side of the relationship, the mechanisms put in place to manage this interface are also influenced by the relative interdependency of the relationship.

The higher the interdependency, the greater their mutual commitment in the relationship, therefore making the companies rely on each other more in order to achieve their objectives, both for the functioning of their supply chain activities and for access to limited resources [LUS 96], which often includes innovation capacity. In addition, the less they need explicit contracts [WIL 75, WIL 85], the more they need elaborate coordination mechanisms [KAL 09, p. 50][12].

On the contrary, when there is an imbalance in the relative dependencies of one another, it manifests itself as an imbalance in the commitment of the two firms [GUN 95]. When interdependency is highly asymmetric, the risks of opportunistic behavior on the part of the most dependent firm are multiplied [PRO 89]. This risk of opportunism, when identified, leads to the strengthening of binding contractual and coordination mechanisms [FRE 04, p. 60]: distribution of patent rights according to expenditure, monitoring of the other's activity, retention of title clauses, termination clauses, etc.

Simultaneously, this imbalance encourages the establishment of unbalanced contractual arrangements and coordination processes that are unlikely to ease the tensions and misunderstandings that result from them.

Thus, through the study of the impact of types of interdependency within a distribution network, Lusch and Brown noted that the least dependent party made the extensive use of formal contracts, thus imposing its standards on the other party [LUS 96].

In the same way, the asymmetry of interdependency leads to a distribution of unequal rights through the formation of contracts that are unbalanced in value and power, to the disadvantage of the most dependent firm [LAS 00]. This then reinforces the asymmetry of interdependency and thus the risk of conflict within the relationship [FRE 04] and opportunism of the most dependent firm [PRO 89].

2.3.1.3. *Familiarity within the two firms*

Familiarity within a business or innovation relationship is the result of the exchange of information between cooperating firms and the creation of knowledge that is specific to the relationship, resulting from its interactions.

12 They can also rely on the involvement of managers on both sides, with the interpersonal relationship allowing the formation of an implicit and non-formal agreement between the two parties.

This familiarity has an impact on the client–supplier partnership's capacity to both cooperate and innovate. The development of each other's knowledge makes it possible to develop the collaborative capacities of each firm, thus improving their ability to benefit from cooperation [SLU 11]. This is the case in both negotiation and execution periods. Greater familiarity enhances the efficiency of negotiations, relationship administration and the division of labor between client and supplier [HÅK 82, p. 30].

Familiarity also contributes to the sharing of a common vision and common relational norms. Familiarity implies that each individual involved in the Open Innovation project has a relatively similar perception of how to interact with others. In the end, this allows them to perceive the potential to be drawn from the exchange and combination of resources [EBA 98] and to find a focal point. This increases both the efficiency and effectiveness of cooperation.

Furthermore, the learning outcomes of pooling resources and skills for innovation purposes represent new information resources generated by the cooperation, which improve each firm's capacity for innovation by expanding its knowledge base [COH 90]. And since this increase in both party's capacity for innovation is one of the reasons for cooperation, its realization contributes to the performance of the relationship.

On the contrary, the overdevelopment of shared relational and cognitive norms also hinders the firms' ability to find new combinations. Too much familiarity leads to redundant knowledge within an inter-firm relationship [COH 90, NOO 99b]. Nooteboom *et al.* specified that there is an optimal cognitive distance to maximize innovation performance between two companies [NOO 07].

Familiarity also leads to a reduced need for control mechanisms within the client–supplier relationship [GUL 95]. As familiarity reduces information asymmetries, there is less need for methods to reduce them. Subsequently, the greater the familiarity, the more the governance mechanisms are adjusted to the particularities of the other and to the specificities of the relationship [DOZ 96].

Lastly, familiarity can also lead to the birth of new innovation projects resulting from to an intimate knowledge of each other's strategies and capacities, etc.

2.3.2. *The four atmospheres of client–supplier Open Innovation and their performance*

These three elements that form the essence of the atmosphere – trust, interdependency and familiarity within the client–supplier relationship – are strongly linked to the performance and governance of client–supplier Open Innovation.

Our quantitative study allowed us to identify four different atmospheres of client–supplier Open Innovation (Figure 2.6) and their links with the different levels of performance.

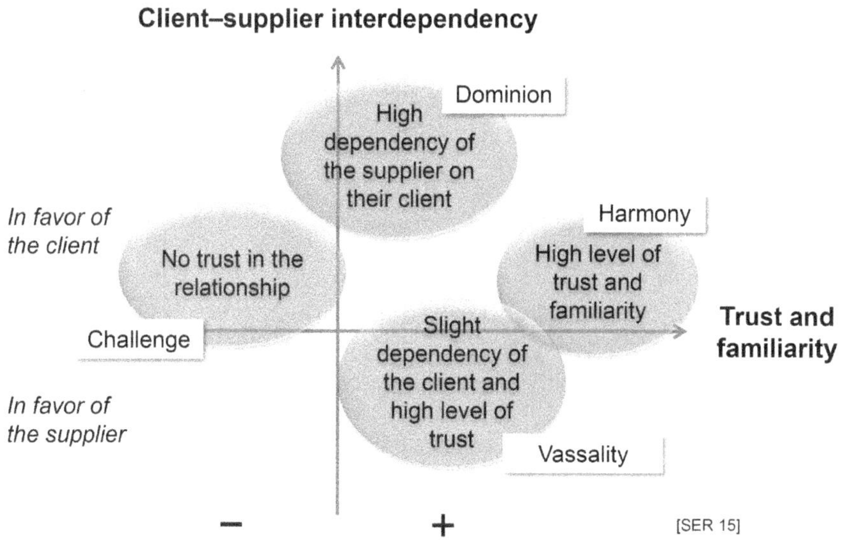

Figure 2.6. *Four atmospheres of the client–supplier Open Innovation relationship*

We can see that the two atmospheres reflecting a poorer quality of the relationship – the one on the left and the one at the top of the figure – barely achieve outperformance, whereas the atmospheres whose quality appears to be the best – the one on the right and the one on the bottom of the dial – are only rarely found in low performance. However, all of them can correspond to achieving performance.

These results reflect both the importance of the influence of the relationship's atmosphere on its performance and its limitations. We will detail them below.

But first, it is interesting to note that the average trust level of our sample is relatively high: 5.45 on a Likert scale ranging from 1–Strongly disagree to 7–Strongly agree[13].

This may seem surprising in a low-trust society such as France [FUK 95, USU 00]. It is less so if we consider that we are interested in inter-firm relations that together lead at least one Open Innovation project (more than 76% of our sample). This means that there must be a minimum of trust between client and supplier. This implies that this trust is, at the very least, based on an obligation of one party to the other [GUL 95] or on a minimum recognition of the supplier's capacity for the targeted innovation, or on the other hand, the client's ability to continue the business relationship.

2.3.2.1. *The "Challenge" atmosphere*

The atmosphere that we have named "Challenge" is characterized by an average level of "neutral" trust. The 34 suppliers who judge their relationship with a client on an Open Innovation project do not disagree or agree with questions relating to if they have a relationship of trust with the other and if the other firm is honest and trustworthy.

In this atmosphere, the level of familiarity, like that of trust, is the lowest of the four atmospheres. If the interdependency asymmetry is equivalent to that of the Harmony atmosphere – i.e. slightly in favor of the client – the levels of dependency of one on the other are much lower, and even lower than in the other two atmospheres.

This atmosphere may lead us to assume that these are relationships that are not very strategic for either of them, and/or that they are companies that are emerging from a closed innovation model – that are struggling to open up to each other.

The analysis of the distribution of types of governance under this atmosphere reveals a near absence of the highest performance. With low

13 With: 5–Moderately agree, 6–Agree.

trust, it is almost impossible to achieve "outperformance"; the relational return is not generated.

On the contrary, the possibility of "achieved performance" (almost half of the sample in this atmosphere) shows that lack of trust is not necessarily an obstacle to achieving the objectives. And that without trust (but no distrust too) and strong dependency on one another, Open Innovation objectives can be achieved, and with a good level of coordination.

2.3.2.2. *The Dominion atmosphere*

The *Dominion* atmosphere is marked by a high interdependency asymmetry, with a strong dependency of the supplier on its client with whom it cooperates on the innovation project. Although the level of familiarity is equivalent to the next two atmospheres, the level of trust is lower.

However, this level of trust is relatively high. This is surprising given the importance of interdependency asymmetry. Indeed, in theory, interdependency asymmetry and trust are difficult to combine.

If the dependency of one party on the other leads to a greater need for trust [LAR 80], situations of unbalanced dependency tend to provoke aggressive and opportunistic behavior on the part of the weakened party, as well as abuses of dominant position on the part of the other [FRE 04], and therefore *a priori* insecure relationships.

This could be explained by the fact that the more a client develops value-generating activities with a supplier and involves it in its innovation projects, the more dependent it becomes on the supplier [KÄK 13].

A first hypothesis explaining this "surprise" would be that, in a context of client–supplier Open Innovation, this is the present-day asymmetry. And the supplier is hoping to limit this asymmetry in the future. Its hope for future rebalancing leads it to want to maintain the relationship and allows it to maintain a relatively high level of trust.

The phenomenon observed is then a mixture of competition and cooperation [DON 08, p. 163] that is reflected in the atmosphere of the relationship. Another complementary hypothesis would be that the clients designated by the suppliers interviewed are dominant clients who do not abuse them[14]. A third hypothesis would be that the supplier's relatively high level of dependency on its client leads to an obligation of results, while limiting overzealousness. None of these three hypotheses are exclusive to the others.

Like the Challenge atmosphere, under this atmosphere, outperformance is almost never achieved. On the contrary, low performance is relatively less present.

The relatively high level of trust allows the limiting of average performance, whereas a strong asymmetry of interdependency does not allow the best performance to be achieved.

2.3.2.3. *The "Vassality" atmosphere*

The "Vassality" atmosphere is the counterpart of the previous atmosphere in terms of client–supplier relations [NOG 02]. It is marked by a higher level of trust, and the asymmetry of interdependency is reversed. This is lower, and slightly in favor of the supplier, whereas the level of familiarity is high.

The distribution of the Vassality atmosphere in all performance levels reinforces the previous result: a higher level of trust leads to a higher level of performance of the relationship, but does not ensure it.

2.3.2.4. *The "Harmony" atmosphere*

The atmosphere called "*Harmony*" has the highest levels of familiarity and trust. The interdependency asymmetry is low, almost balanced, although slightly in favor of the client.

This atmosphere is the one where "low performance" is rare. It thus confirms best practice studies that show that an atmosphere of high trust and familiarity corresponds to a relationship where relational return is also high [DWY 87, KAL 09, LEE 12]. With a balance of dependency, trust and high familiarity, the performance of client–supplier Open Innovation is likely to meet or exceed everyone's expectations.

---

14 This would probably constitute a bias in our study, unless we consider that a majority of France-based client companies do not abuse their dominant position in similar cases.

2.3.3. *Which atmospheres correspond to which type of Open Innovation governance?*

Examining the link between atmosphere and governance reveals a strong predicament.

We can see that in the two atmospheres reflecting a poorer quality of the relationship – Challenge and Dominion, those of the left and the top of Figure 2.6 – the most adopted type of governance is "Unstructured" governance. If "R&D project" governance is not very present in the Dominion atmosphere, then "Exclusive partnership" governance is the one that will not be very present in the Challenge atmosphere. Moreover, in the atmospheres whose quality appears to be the best – Dependency and Harmony – the different types of governance are distributed roughly equally.

	Unstructured *Cluster 1*	**Developed partnership** *Cluster 2*	**R&D project** *Cluster 3*	**Exclusive partership** *Cluster 4*	
Vassality atmosphere	34% *16* **27%**	17% *8* **21%**	23% *11* **33%**	26% *12* **41%**	100%
Dominion atmosphere	56% *14* **23%**	24% *6* **16%**	8% *2* **6%**	12% *3* **10%**	100%
Harmony atmosphere	25% *13* **22%**	34% *18* **47%**	19% *10* **30%**	23% *12* **41%**	100%
Challenge atmosphere	49% *17* **28%**	17% *6* **16%**	29% *10* **30%**	6% *2* **7%**	100%
	100%	100%	100%	100%	

% in row *Sample size* **% in column**

[SER 15]

Figure 2.7. *Distribution of forms of governance according to the atmospheres of the relationship*

2.3.3.1. *"Unstructured" governance atmospheres*

It appears that Unstructured governance is found relatively less present in the atmospheres of the highest relational quality: that of Harmony and Vassality.

Our hypothesis is that, in a Challenge atmosphere, governance influences the atmosphere: the absence of formal relational mechanisms such as contractual clauses specific to the innovation project does not allow the laying of foundations for a strong level of trust.

And under the Dominion atmosphere, it is the atmosphere that influences governance: the asymmetry of interdependency leads to the implementation of a governance system that favors the (asymmetrical) client–supplier relationship, or the absence of mechanisms specific to the innovation project, that are also specific to relationship management.

2.3.3.2. *"R&D project" governance atmospheres*

Similarly, the rarity of cases which have adopted R&D project governance in a Dominion atmosphere seems to indicate that the latter, focused on the upstream stages of innovation (presence of R&D-oriented methods and absence of methods relating to the client–supplier relationship), are hardly compatible with a strong asymmetry of interdependency in favor of the client.

One possible explanation would be that suppliers in a weak position refuse to take part in an R&D project for which there are no resale rights that can counterbalance its dependency. This phenomenon is reflected in the American automotive industry, where manufacturers' suppliers primarily seek to develop innovation projects with those they are not dependent on [SCH 06, p. 930].

Simultaneously, the "imposed and balanced" formalities of this type of configuration would reinforce this explanation: the projects described are mainly collaborative projects that are institutionally funded – in other words, according to the standards required to obtain these subsidies. In this case, however, the least dependent party, the client, would prefer to opt for its own standards or the absence of these standards, with the Unstructured configuration corresponding to a contract that is more easily negotiable and therefore more favorable to it. Another explanation would be the difficulty of clients to consider their "dependent" suppliers as innovative firms to whom they can entrust pure innovation projects.

2.3.3.3. *"Exclusive partnership" governance atmospheres*

The Exclusive partnership configuration is found almost exclusively in the Harmony and Vassality atmospheres, which are those with the highest levels of trust and familiarity. The important contractual link of this configuration, which is an exclusive purchasing commitment in the event of success combined with reminders of the contract as a strategy of influence on the part of the client (combined with incentives), favors this type of atmosphere and vice versa. Trust and the contractual and relational mechanisms of control of a relationship complement and influence each other [DAS 98, DWY 87, GUL 95, HAU 10, MAC 80].

2.3.3.4. *"Developed partnership" governance atmospheres*

Developed partnership governance is found in all atmospheres, and is being particularly present in the Harmony atmosphere (47% of Developed partnership projects for 34% of the Harmony atmosphere). Compared to the Exclusive partnership, the inter-firm link is not so much the result of exclusivity but rather the variety of elements that make up the governance of the relationship.

The notable difference is that the Developed partnership includes elements involved in safeguarding and controlling the relationship, based on a fear of opportunism and deviation of the other: complex contract, limiting supplier information asymmetries through the use of external experts, reminder of the contract and use of threats by the client. Here, we see the observations of Hausman and Johnston [HAU 2010] on the positive influence of a minimal coercive attitude on inter-firm trust.

A final surprising point of this type of governance is finding a link with the Dominion and Challenge atmospheres. Our hypothesis is that this configuration offers the possibility of balancing interdependency for the former and the conditions of trust for the latter: the spectrum of future orders allows the supplier to engage in the project, without balancing the current atmosphere of the relationship, unlike in the case of the R&D project configuration.

This then raises questions about the immediate impact of this type of fair sharing of the fruits of the joint project on the commitment of the parties in a relationship [RIN 94]: it can only be considered in a dynamic perspective.

2.4. The innovation project and its influence on governance

2.4.1. *Innovation maturation and the evolution of governance*

An innovation project generally follows four main phases that reflect the increasing maturity of the project from the new idea to its launch on the market: the idea generation phase, the planning phase, the execution phase and the transfer to industrialization phase [COO 90, VAN 14, WHE 92].

As the innovation project progresses, both the activities and the players involved evolve in the firms [JOH 11]. The expectations of firms, the dyadic players and the division of tasks within the client–supplier partnership gradually change [LED 11, SÄF 14, VAN 14].

Some phases of the innovation project and the nature of the activities that are specific to these phases may favor the co-location or separation of the project players from both parties [LAK 06a]. In the project definition and launch phases, it is important that client and supplier teams share the same place for creative and responsive exchanges. Between the two, coordination and exchanges are more effective when they are done in a more punctual and focused manner.

Contractual arrangements are also differentiated according to the level of maturity of the innovation [NPV 09]. Nestlé's partnership model thus favored certain contractual arrangements, and certain types of suppliers according to the maturity level of the innovation project [TRA 09].

This differentiation of contractual arrangements according to the maturity of the innovation comes from the fact that, upstream of innovation projects, for a client, the project is at its most flexible and there are many possibilities to make choices [MID 98]. Then, as the project progresses, the level of knowledge about it increases. This corresponds to a decrease in uncertainties, and also in the capacity for action and degrees of freedom in the design of innovation. In addition to this, it is combined with an increase in costs incurred and modification costs [MID 93, p. 132, WYN 98, p. 41].

To this end, it is important to collect a lot of information from the early stages of the project in order to assess the feasibility of the markets and technologies contemplated, but without seeking commitments. The client will then seek performance rather than cost [UTT 75]. And even further upstream, the client will seek to explore and/or create new knowledge and innovation potential.

Thus, for the most upstream phases of innovation, Sadeh, Devir and Shenhar [SAD 00] showed that the performance of R&D contracts is better when the client agrees to cover research costs rather than pay a fixed price – the commitment of the other party is more assured when it is assured of the "non-loss" of a requested work, on which there is a high degree of uncertainty with regard to the efforts to be produced.

From the client's point of view, early integration of a single supplier involves a risk, one that involves the supplier leading to a lock-in situation by favoring the choice of technologies that correspond to its own resources – which, in addition to limiting the possibilities of choosing solutions for the project, increases the future risks of hold-ups, and thus the implementation of appropriate backup mechanisms such as exit clauses or facilitating the possibility of a second source.

Moreover, involvement of suppliers in the downstream phases of innovation of management mechanisms that are poorly articulated with a sharing of "white box" responsibilities improves the efficiency of the project, but not learning opportunities [SOB 02]. And conversely, for involvement in the upstream phases, developed information exchange mechanisms and "black box" or "open box" sharing generate higher coordination costs and provide an opportunity to effectively access external knowledge that can then be used in other projects.

Sometimes the contractual arrangements are also revised at each successive phase of the project [CAL 13, MID 00], whether planned from the outset or modified after the event.

A first strategy is to amend the contract that was initially established. A second strategy is to set up additional contracts. A third strategy is to set up modification or iteration protocols based on an initial contractual arrangement that can be extended according to the progress of the projects.

The three strategies can also be combined. Consequently, the governance of client–supplier Open Innovation is adapted to the maturity of the joint innovation project, and must be adapted as it matures.

2.4.2. *Which type of governances corresponds to which stages of maturity?*

Our sample of client–supplier Open Innovation cases reveals that Unstructured and Developed partnership governance are found at all stages of innovation maturity (Figure 2.8).

R&D project governance is almost exclusively selected when the supplier is involved from the earliest stages of the innovation project. Exclusive partnership governance is basically not selected for involvement in the industrialization phases.

	Unstructured *Cluster 1*	**Developed partnership** *Cluster 2*	**R&D project** *Cluster 3*	**Exclusive partership** *Cluster 4*	
Upstream	35% *35* **58%**	16% *16* **42%**	30% *30* **91%**	20% *20* **69%**	100%
Design prototyping	34% *10* **17%**	38% *11* **29%**	3% *1* **3%**	24% *7* **24%**	100%
Industrialization	50% *15* **25%**	37% *11* **29%**	7% *2* **6%**	7% *2* **7%**	100%
	100%	**100%**	**100%**	**100%**	

% in row	*Sample size* **% in column**

[SER 15]

Figure 2.8. *Distribution of forms of governance according to the maturity of the innovation*

This can be explained by the fact that these two types of governance are particularly characterized by a relatively lower presence of purchase–sales entities compared to R&D entities.

The link between R&D project governance and the low maturity of the innovation project is obvious. On the contrary, the weak link between the Exclusive partnership configuration and the downstream phases of innovation is surprising, whereas this is not the case for Developed partnership governance.

This result may be explained by the small sample size and non-response bias. Another hypothesis is that downstream of innovation, when there is a high probability of failure, a firm may more easily consider entering into an exclusive contract – this is sometimes the only way to convince a partner, whether client or supplier, to participate in a joint project that is subject to high uncertainties [RIN 94]. On the contrary, on a timeline where the uncertainties of innovation are relatively lower and the chances of success are perceived as being higher, exclusiveness appears to be too high a debt to pay.

2.5. Forms of governance to be favored according to the atmosphere of the relationship and the maturity of the innovation project

By integrating all the links described above, which have been derived from academic theories and observed in practice, we can provide a first answer to the question of how to involve clients and suppliers in an Open Innovation project.

First of all, whether you are a client or a supplier, you must be aware that in order to get the best performance, you have to consider:

(1) all interactions, between the two firms and within each firm;

(2) the atmosphere of the relationship;

(3) the progress of the innovation project (its technology readiness level or its progress toward time-to-market, depending on the culture of the firm).

Some types of governance that we have identified and described allow us to get the best performance from the relationship. In order to do this, they must be adapted to the atmosphere of the relationship and the maturity of the innovation project at the point in time when the client and supplier together commit themselves to the Open Innovation project.

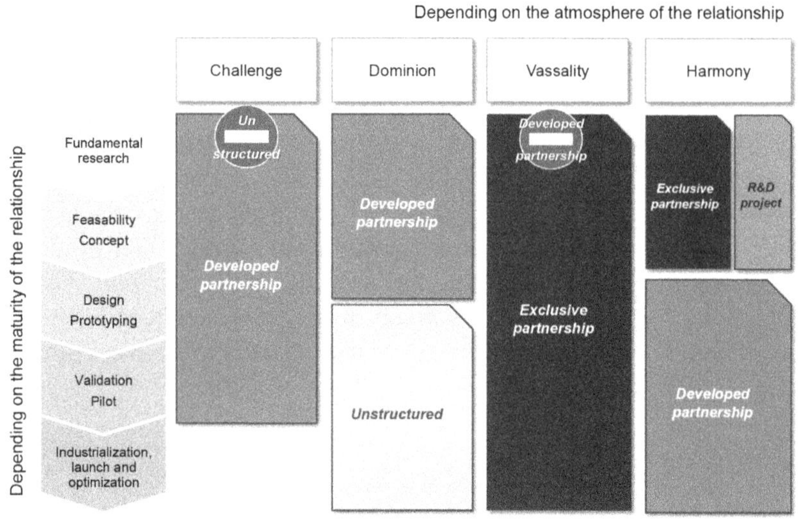

Figure 2.9. *Which governance should be favored for client–supplier Open Innovation? For a color version of this figure, see www.iste.co.uk/servajean/dynamics.zip*

In a *Challenge* atmosphere, where trust and familiarity are lowest and there are few mutual dependencies, a relationship must be built. First of all, Unstructured governance should be prohibited for the fundamental research phases. Then, it will be the Developed partnership that should be favored. Thanks to developed inter-firm coordination and resale rights provided for if the cooperation is successful, the conditions for a successful relationship will be in place.

In a *Dominion* atmosphere, when the supplier depends very heavily on its client, for the upstream phases of the project, a Developed partnership governance is the solution that provides the most performance. The importance of the commitments offsets the power asymmetry between the two firms. For the design–prototyping phase, this governance can be a second choice, after Unstructured governance.

Then for the more downstream phases, Unstructured governance allows for better performance than other types of governance. The client then has to adopt a highly transactional approach, close to the "life cycle" approach and the client–supplier business relationship. It should be noted that while this form of governance provides better performance than others – it is effective and efficient in this, and only this, context – it does not allow outperformance to be achieved.

In a *Vassality* atmosphere, for the phases of fundamental research, Developed partnership governance is to be avoided, whereas Exclusive partnership governance is to be favored regardless of when the supplier is involved with the client. The option of a relative openness of the results to competition of upstream Open Innovation seems counterproductive when the client is dependent on its supplier.

While the contractual arrangements of the Exclusive partnership make it possible to encourage the creation of such an atmosphere (the client is mainly committed to buying the result of the joint innovation project from the supplier), the cumulative effect of the contractual arrangements on the atmosphere nevertheless carries the risk of reinforcing the asymmetry of interdependency in favor of the supplier. Furthermore, regarding the question of defining the configuration of the partnership in a Vassality atmosphere, regardless of the maturity of the innovation project, if the Developed partnership seems to be prohibited according to our results, it will be recommended to managers that they favor the implementation of the management mechanisms of the Exclusive partnership. Then, managers will be able to look at the design of the contract so as not to encourage too big an imbalance that can lead the client into a situation of confinement/ dependency that is too great, and therefore seek to withdraw as a result.

In a *Harmony* atmosphere, for the upstream phases of innovation, R&D project and Exclusive partnership governance are to be favored. This means clearly sharing roles, responsibilities and results, and jointly monitoring the project's progress. In a context of strong uncertainties about innovation, the quality of the atmosphere makes it possible to adopt two extreme forms of governance. Without any commitment following the project there is no need to be motivated themselves contractually beyond the established framework. Or with exclusive commitment, alternatives for both are not considered – the relational return is in full swing and is maintained.

Then, for the downstream phases, from the prototyping phase to industrialization, Developed partnership governance appears to be the best choice. As the market approaches, there is a need for additional control through contractual and coordination methods. In order to maintain a balance in the relationship, the exclusiveness of resale rights gives way to a commitment to minimum volumes.

Based on this table, the performance of ongoing relationships can be improved, if necessary, by seeking to gradually and simultaneously adapt governance.

The first recommendation is to work on management methods before tackling contractual arrangements.

It is then a question of trying to encourage the implementation of resale rights in contracts governing innovation projects. This requires going beyond the sharing of intellectual property and considering the possibilities of sharing the value created, by reflecting on possible economic models for the duo. This can then lead to resale rights that can range from the valuation of shared intellectual property to the client's commitment to order from the supplier and the supplier to deliver to the client, in the event of success.

The second recommendation is based on the discovery of the importance of client proactivity for relationship performance. Although proactivity is measured here as a performance indicator, it can also become a governance mechanism for the relationship. In order to do this, all that needs to be done is to institutionalize proactive behavior with suppliers.

What is often criticized, particularly in comparisons between France and Germany, is the lack of client invitations to their ecosystems for new calls for tenders. For example, giving regular information to your supplier and especially informing it of new quotation requests in progress is a factor that contributes to the construction of performance. It can also mean quick answers to questions or spontaneous suggestions for improvement (a lighter, less formal version of the value analysis).

These are all mechanisms with relatively low implementation costs, but their impact on performance is high. However, this involves prioritizing innovative suppliers and therefore raises the question of whether it is appropriate to manage a panel of innovative suppliers in a similar, but potentially distinct, way to a panel of strategic suppliers.

3

The Dynamics of Client–supplier Innovation Interactions – History and Analysis of a Collaboration

In spring 2012, the DELTA project is launched. The DELTA system is a technological innovation which has to bring new functions to the system in which it is integrated. In only 20 months, the DELTA system will have to be designed, approved and presented by DELCAR to an automobile manufacturer.

However, to create this system, there will be numerous challenges. A DELTA system of such complexity has never been achieved. It requires the use of new material of which the behavior is not fully known. The methods of integration of classic parts in such a system are not defined. The assembling of these can only be done together with the manufacturing of the main part. This assembling process has never been carried out in an automobile environment and with its related constraints. The manufacturing cost is still no more than a target.

DELCAR cannot overcome these challenges alone. It is necessary for them to find an external partner who knew how to work with them on this innovation project. Then, by working with them, they might overcome all the challenges.

In this chapter, after having outlined the context of the DELTA project, we will follow the relationship between DELCAR and its supplier week by week. By adopting the same interpretation framework as before, we will explore *in vivo* the dynamics of Open Innovation between the two companies. Then, we will highlight what this case and this interpretation has taught us about the dynamics of client–supplier Open Innovation.

3.1. The context of the collaboration

3.1.1. *DELCAR: an automotive equipment division*

DELCAR is one of the main divisions of one of the leaders in automotive equipment. With a sales revenue of several billion euros and tens of thousands of employees across the world, DELCAR designs and manufactures key systems for civil automotive vehicles.

Within the supportive functions of DELCAR, there is "Research and Development" and "Purchasing" (Figure 3.1) – the main functions that we monitor in this research.

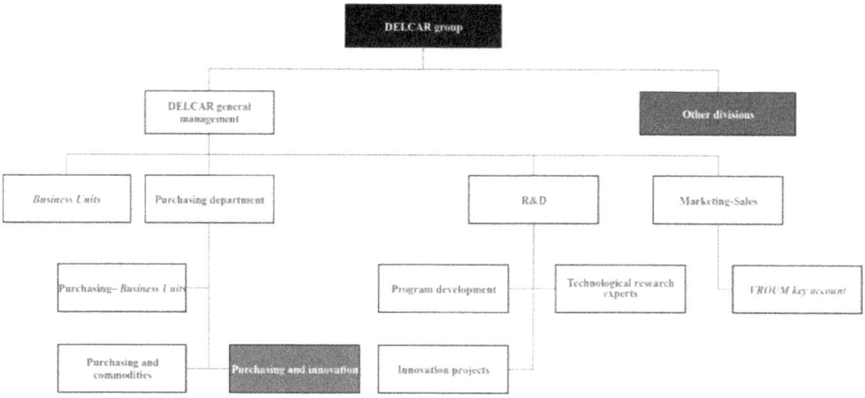

Figure 3.1. *Simplified organigram of DELCAR. For a color version of this figure, see www.iste.co.uk/servajean/dynamics.zip*

Within R&D departments, there is:

– the Technological Research department, bringing together experts who are responsible for supporting DELCAR's technological capacity;

– the Department for Innovation projects, responsible for the generation and management of innovation projects until they are passed over to the Program Development department;

– the Program Development department, responsible for preparing and launching the industrialization of programs, which are the subject of a commercial contract signed for a particular vehicle.

The "Innovation Projects team" is found on two Innovation floors, large open spaces dedicated to innovation projects, in two main R&D Offices, based in Europe, with a global operational scope.

The Purchasing department is split into three entities:

– Commodity Purchasing, responsible for the purchasing of commodities common to all Business Units (metal, plastic, etc.);

– "Series Production" Purchasing, linked specifically to each Business Unit and each program/vehicle platform;

– the Innovation-Purchasing team, which is the studied entity, in connection with the "Innovation Projects" teams.

Finally, within the Sales and Marketing department, there are teams dedicated to each automotive manufacturer, including the firm VROUM.

The observed entity: the Innovation-Purchasing team

The Innovation-Purchasing department, an entity which provides functional support to "Purchasing", has the role of liaising between "Purchasing" and "Research and Development" and ensuring the performance of DELCAR's co-innovation projects that are not conducted with clients. Its operational scope is global.

It is currently made up (at the time of writing) of the Innovation-Purchasing Director and a Manager. The former is based between the firm's Head Office and the R&D Office, where the innovation collaboration described here happens, at the same place where half the Innovation Projects staff are located. The latter is based in the other R&D Office, with the other Innovation Project teams.

Members of the Innovation-Purchasing department are responsible on the one hand for researching and selecting innovative suppliers and external innovative ideas for the organization. They are also responsible for helping the R&D teams and organizations outside of DELCAR in the implementation and the supervision of innovation cooperation.

3.1.2. *The Innovation Project at the heart of the relationship: the DELTA project*

The DELTA project has the goal of creating an object that had never been created before and of which the industrial feasibility is uncertain: the DELCAR system but with a new design, which requires the insertion of joint parts manufactured from solid metal and the use of a composite material never used in these conditions: material α. The DELTA system therefore consists of a DELTA part made of material α and in which the metallic inserts G and H are integrated.

This is an innovation project from DELCAR's point of view. This innovative system might reduce the total weight of the DELCAR system by approximately 20%. So, it responds to the requirement for a reduction in the weight of vehicles. It might also allow them to propose new possible shapes to automotive manufacturers and an increase in space to passengers by reducing the thickness of the system by 15%.

Innovation management at DELCAR

Innovation at DELCAR is a "*New and Innovative Product development ahead of a client awarded program*", as defined by the internal process of innovation management. It relates to every new product, service, technology or manufacturing process. The innovation projects are managed following a stage-gate process [COO 90] of five phases, from ideation to transfer to development programs (Figure 3.2).

Gate number	Title	Decisions at the gate
0	Project objectives	Project start Choice: single or generics mode
1	Concept selection	Statement of work signed during the gate Concept selection
2	Concept validation	Go to gate 3 validation with the concept
3	Product and process validation	Product / process allowed to be incorporated into a product offer
4	Transfer or additional validation	Product / process on shelf

Figure 3.2. *The five phases of the process of managing innovation projects at DELCAR*

When a project is launched (phase 0 passed), a Project Manager belonging to the "Innovation Projects" department is appointed and brings together a project-coordination team.

The other members are split functionally and physically into their respective departments; they generally come from the Technical Research department. They are brought together in a common space within one of the two innovation floors.

The Project Manager is in charge of coordinating their activities and following the progress of the project and its budget. He/she has little influence on the individual evaluation of the project team members. The technical decisions are carried out under the leadership of the architect connected to the project whose role is to coordinate the design activities; such decisions are approved at each passed gate.

Phase transition, the passage of gates, takes place during a monthly innovation committee meeting organized by the Innovation Director, who is directly responsible for innovation projects. This committee systematically brings together the n-1 of the Director of Innovation and the Directors of Innovation-Purchasing, of Marketing and of Quality. The Innovation Project Managers and the Heads of the technical domains of the Technical Research department are present when the subjects concern them.

Some structural elements of this system that show similarities have already been created successfully by two expert teams from the Technological Expertise department at DELCAR but with a lesser degree of complexity and without going beyond the prototype stage.

If the main manufacturing technology envisaged is known and largely used in the automobile sector, its implementation with material α is not controlled because of its anisotropy and the difficulty of ensuring its homogeneity.

Today, it is impossible to numerically model the behavior of such material, notably those relative to the constraints required by the automotive standards such as the behavior during crash tests. Also, when there are other elements which are added to this material with diverse mechanical properties and relationships between them which are not completely controlled, such as the insert of elements made with other materials (inserts G and H), or the deposition of pigment to color it, the uncertainty is even greater.

The benefits and the limitations of the material alpha in relation to alternative solutions

The materials used in the structures of automotive equipment need to combine both a strong resistance and elasticity. They must be ductile during the crash test.

As for its density, type V-W metal – at the heart of DELCAR's business – is an ideal material for the structures of a given automotive vehicle: a strong modulus, a strong resistance and a malleability which prevents failures during the crash test. The respective resistances and elasticities of the V-W metal and its alternatives (Y metal and others) are similar.

The π composite material implemented in other industrial sectors possesses, at best, a resistance equivalent to that of traditional material. If its modulus of elasticity is low, it can be increased through the design of certain aspects of the targeted part (by adding ribbing, by defining the variable layers etc.). The main limitation of this type of material is its limited elongation before breakage and its prohibitive price.

On the other hand, composite material α has a strong resistance, a medium modulus of elasticity and a high impact resistance. This makes it a strong candidate for the interior parts of automotive vehicles. If it is cheaper than the π composite material, it is also less resistant in some conditions.

The implementation and the maintenance of these inserts in manufacturing molds are difficult techniques, as there exist no proven solutions respecting automotive cycle time. The feasibility of this crucial step is not fully known. For the Manufacturing Process Manager at Delta, "*It's a new process, we need to gain experience*" and as for the Tool Purchasing Specialist "*These inserts will never hold*".

The joining inserts pose another problem: they must ensure that the central part of the system is held together by a strong link with material α and must ensure that the sealing of the joint zones cannot be penetrated by the injected material (difficulty point G).

The collaboration with a mold supplier must be able to solve this technical difficulty: "*the selected supplier must be able to start helping us with the design of (these) zones.... The sizing of the (system) is entirely linked to these zones and we do not have the skills internally to progress much further than we already have*" – Manufacturing Process Manager.

Finally, the experts from the Technology department have highlighted three lever points allowing the success of this type of part for automobile series production:

– the smoothness and automation of the production process, making it possible to obtain parts which have a price level equivalent to those from proven technologies and optimized processes;

– the mastering of the part's design and its ability to respond to automobile standards, firstly those linked to crash tests; the modeling of the behavior of such a system is limited nowadays;

– the manufacturing cycle time of the part with the integration of the inserts.

The goal of the DELTA project is to validate the feasibility of this system through the design and production of object-prototypes, which will then be tested: qualitative evaluations, crash tests, capabilities, etc.

The results of the two preceding DELTA-type projects did not allow projections on these different feasibility elements: "*in the crash, the rear part of the system exploded*". Furthermore, DELCAR's industrial tool is not able to produce this part – DELTA's technology system was concurrent with DELCAR's main technology. On the other hand, another division of the group possesses an industrial tool which corresponds to this technology, with other materials and without the insertion of joint parts. Finally, the procurement of material α is not guaranteed: the procurement options, like the purchase price, are dubiously estimated.

The financing of this project is shared between DELCAR and an automotive manufacturer (VROUM) interested in the features offered, provided the proposed solution is at the same cost as the existing solution for a targeted vehicle range (VR2020). The results of this project have to be included in the planned launch of VROUM's new products: if the results are positive, the DELTA system will have to be submitted for quotation of DELTA type systems for VR2020 within 20 months.

The main people involved in the project were, from DELCAR's side, from the DELTA project team: 7 Engineers and 1 Quality Manager. They share the same work island on one of DELCAR's Innovation floors.

The coordination within this team is carried out by holding a daily meeting lasting 5 minutes, and every Monday, an hour-long meeting which takes the form of a debriefing. If the formation of this project team can in theory evolve over the course of the project according to its needs and its functions, the DELTA project-team will not evolve during the course of the 20 months.

The DELTA project-team is made up of:

– Zeus, *The Project Manager.* He manages the project and functionally oversaw the other seven members of the project team. He reports to the Innovation Floor Manager. His role is to ensure the coordination of different resources and to lead the project through an innovation management process until it is passed over to the development teams who will prepare the market launch. He is also responsible for maintaining the relationship with the car manufacturer VROUM's VR2020 team, which sometimes requires him to spend an afternoon a week at their office;

– Hades, *The Senior Architect.* A very experienced engineer and an expert in DECLAR's complete system and its structure, he coordinates the design activities;

– Mario, *The Manufacturing Process Manager*, and Luigi, *his assistant.* They report directly to the Head of Manufacturing Processes of DELCAR's system structures – himself belonging to the Program Development department. They approve the process that allows the manufacturing of the product designed by the other team members. Mario is responsible for the product-process design coordination: "*my job is to get everyone in the same room*". He is considered a specialist in material α implementation. Throughout the length of the project, he has to maintain the technical interface with the supplier and has to remain the technical contact for them. His assistant specializes more in the implementation of classic/metal materials;

– *three Engineer and Industrial Designers* for the computer-aided design of the different parts of the DELTA system;

– *the Quality Manager* who ensures that the different "*work packages*" defined in the process of innovation management and that are specified for the project are properly implemented and compliant. She is responsible for analyzing and monitoring the risks at each stage of the project.

This team is supported by other people from different departments within DELCAR. A computing center in another country carries out the project's calculations. A team of four experts specializing in the implementation of materials, products and processes – belonging to the Technological Research Direction – can be found at the other end of the Innovation floor. Each of them can be called upon to help with any problems. Dionysus, *the Material Expert*, the person most involved in the DELTA project, is specialized in mold design.

Experts from another division of the DELAUTO group work on processes close to the project's goal. They are called upon when the industrialization of the production of DELTA parts is addressed. The Manager in charge of innovation within the VROUM Marketing and Commercial team maintains the relationship between the DELTA project team, the DELCAR sales team and the entities involved at VROUM.

The Innovation-Purchasing department contributes to the project through the involvement of the Director of Innovation-Purchasing, Orion, and a member of his team, Asclepius, the researcher occupying the role of Innovation-Purchaser and author of this book. To contribute to the project, the Innovation-Purchasing department is also asked to mobilize a buyer specialist for the purchasing of serial-life tools close to those allowing the implementation of the α material.

3.2. History of a client–supplier innovation collaboration

3.2.1. *Engaging the relationship*

End of spring 2012

The DELTA project passes stage M1: the technical concepts are selected. To reduce the weight by 20% and open up the possibilities of innovative design, material α will be used and the main innovation principles are defined. There will be G and H inserts to ensure articulation and support, respectively, and crash safety.

Nevertheless, they have to prepare for the approval phase with the intention of putting forward a sale proposal to VROUM in Spring 2014: the product and the manufacturing process also needs to be approved. This phase of the project has to allow validation of DELCAR's ability to master the manufacturing process. The year 2018 will then be the date of the first serial manufacturing if VROUM wishes to install this innovative system in the VR2020 vehicle. To this end, in less than two years, they have to present a functionally, a technically and economically approved "on-the-shelf product" to VROUM.

The project-team is built around Zeus. The project schedule foresees that, within one year, the equipment will be launched to test the production of the part. During the project, Mario, the Manufacturing Process Manager, is in charge of the subject. He is supported by Luigi.

In the months before, on another project involving the same team, on another type of system, a supplier was involved without involving Purchasing in the selection and negotiation process. Costs slipped to the dismay of the Project Manager: "*this supplier had an object to achieve. We gave him the specifications. We made a mistake and asked them to make changes. He billed us for a high price... we had no buyer, it's #@!, we were taken for a ride*".

Also, he is asking for support from the Innovation-Purchasing department for the sourcing of the supplier who will carry out this equipment. It is believed that the supplier will be selected in January 2013.

There are no resources available within the Innovation-Purchasing department.

August 2012

Orion, Director of Innovation-Purchasing, directs Mario to Eddy. Eddy is a Buyer specializing in the procurement of molds for the manufacturing technology of main pieces for other parts of the DELCAR system. Normally, he only intervenes for the purpose of mold procurement, which will be used to manufacture series parts.

Mario presents him with specifications: functional specifications, the photo of a prototype close to what is wanted, the elements that constitute the DELTA system, the schematic target process, the list of what the supplier will have to provide (creation of the mold, complete parts and quality control means, support for the design and testing of the parts) and the targeted delivery date.

For Eddy, what is requested does not exist in series. To make mold makers work on the project, it will be necessary to "*bring them something solid and promise them business… If not, I can kiss goodbye all my operations*".

As a result of this exchange, neither of them get back to the other. And, later, Eddy will comment that "*The demand was too low. I did not even advertise it. We have to make a square file and tell them what it pays... this first specification was too vague, I asked for clarification and I did not move while waiting for them*".

Mario then plans consultations with five suppliers including MixMat.

MixMat is considered a specialist in complex mold design and manufacturing for the main material α implementation technology in other industrial sectors. Three other suppliers are recommended by manufacturers from the automobile sector. The final one is another division of the group belonging to DELCAR and has an industrial tool compatible with this technology; it can eventually produce the innovative DELTA system.

The Innovation-Purchasing Director has these four external organizations sign a confidentiality agreement for six years on "*development of a prototype tool on the type of product from the DELTA project and the manufacturing of parts*".

The supplier: MIXMAT, an expert SME in its field

MixMat is a firm that designs and manufactures equipment and molds for a large variety of technologies. They are experts in the design and manufacturing of tools and molds that allow material α implementation, a key element of the DELTA project, but also for the field of aeronautics.

A long-standing market leader, MixMat, is an SME with around one hundred employees. It was taken over 10 years ago by its current president, after a legal redress provoked by the relocation of automobile manufacturers. After regaining profitability for four years, with a portfolio of over 100 clients, MixMat devotes 10% of its turnover to R&D.

MixMat is involved in multiple collaborative innovation projects, some of them involving the group DELAUTO (of which DELCAR is a division). Active in several industrial sectors, automobiles are their new development focus.

Given its novelty, the DELTA project would be for them a first reference for the automobile field within the design and manufacturing of complex molds.

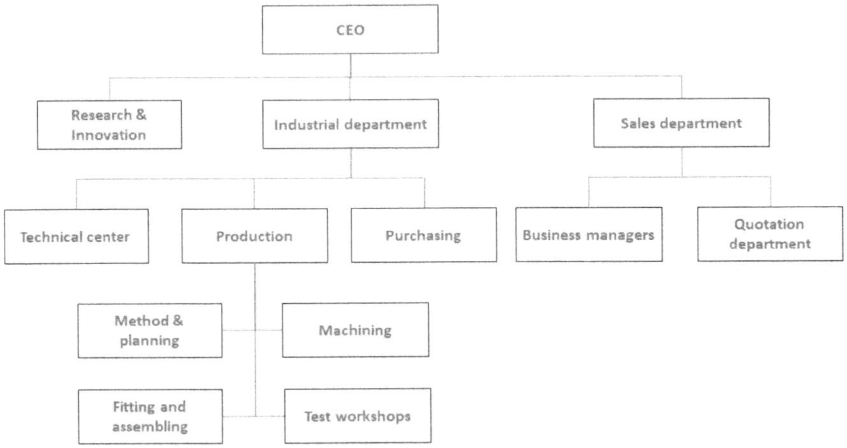

Figure 3.3. *Simplified organigram of MixMat*

In the projects led within MixMat, a Business Manager for the Sales department (Figure 3.3) is named and follows each project "from A to Z" by remaining the key point of contact for the client on the large majority of subjects covered; "*It's the way we work. You have to be able to wear all the hats*".

He calls upon different people according to the phase of the project: "*For example, when we have a small study to approve a kinematic, it's the Quotation department. When the project starts, we can finally introduce the Technical Center. We explain to them how it was quoted... and they make their analysis*".

The idea is not to "disturb" the engineers at the Technical Center until the actual start of the projects – "*It's how we work*". The orders are then carried out by the Industrial Management teams.

For the DELTA project, Mercury was the selected Business Manager. In their relationship with DELCAR, two other people are involved: Apollo, the Commercial Director, and Vulcan, the Technical Manager.

3.2.1.1. *Analysis: the generation of cooperation – context and means*

In this first episode, the first exchanges between the two companies began. Both of the companies discovered the existence of the other from an innovation viewpoint. Both firms considered the other as a potential partner.

The initiating event was the passage of a project milestone at DELCAR. The scope of the DELTA project was fixed. The sourcing of the joint-innovation partner began.

3.2.1.1.1. The atmosphere of the relationship

Before this event, the majority of people who would be involved in the relationship were already aware of the other firm: DELCAR is one of the current leaders in its field, while MixMat is also a leader in its own field, ever since changing its name. The two companies are also taking part in the same collaborative project in a close research area, but that is emerging at the moment and does not involve the same counterparts.

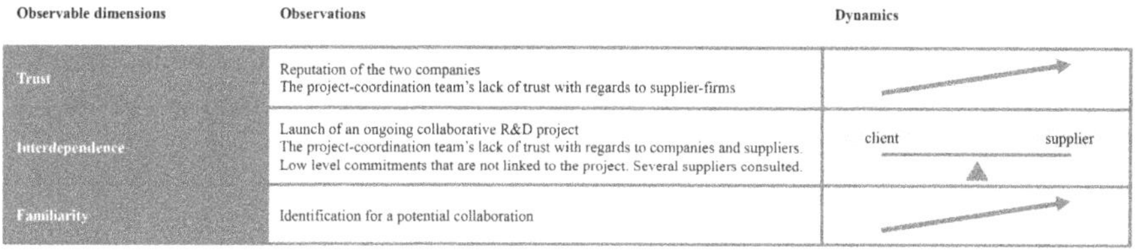

Figure 3.4. *Relationship atmosphere during the discovery stage*

Beyond each firm considering the reputation of the other (Figure 3.3), the level of trust of the DELTA project-team with regard to companies and suppliers is low. This is due to their painful collective experience with a former supplier.

Also, for this project, the Innovation Purchasing department has actually been requested, in accordance with DELCAR's processes, to get involved in the project, in order to select suppliers and contractualize the relationship.

3.2.1.1.2. Maturity of the innovation project

The DELTA innovation project is considered as a new product concept, requiring a new manufacturing process, one going hand in hand with the other.

DELCAR is aware of the difficulties linked to the DELTA system: its integration with the automobile, the test requirements (crash tests) and the manufacturing constraints (quality, cost and acceptable cycle-time). Nevertheless, it is ignored whether this system would be able to meet such requirements: the manufacturing and assembling process is not clearly defined, the modeling and simulation tools are not able to predict the system's behavior, and financial modeling has not been carried out.

The central part's design is being defined; certain zones cannot be defined without exchanges with the supplier who will know how to make or will make a mold which will be used to make this part. According to the Process Manager: "*We need to design the part according to the mold*".

Finally, if this project is carried out for an automobile manufacturer, there is no guarantee that it will then go into development, or that orders will result from it.

From their side, MixMat has knowledge of designing molds for parts made from material α, or its equivalent, and integrating inserts. Nevertheless, this knowledge does not apply to the automobile sector, for large series or high production parts, or with very complex articulated inserts.

3.2.1.1.3. Management of the relationship

Within the DELCAR extended project team, resources are designated to source a supplier who will (1) make the necessary mold for the prototypes, (2) carry out tests following the new process and (3) assist DELCAR in finalizing the definition of its new product and manufacturing process. This is a technical resource on the project-team's side: the engineer in charge of defining the new process, the Process Manager and another resource on the Innovation-Purchasing department's side. On this side, due to a lack of available resources, an internal expert in the Purchasing department is put forward as an expert-buyer to consult (Figure 3.5).

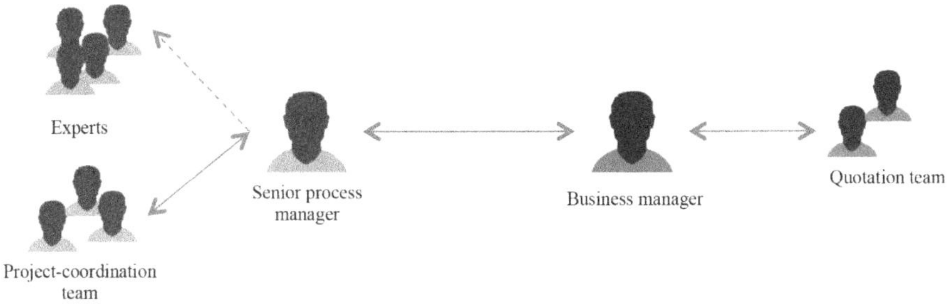

Figure 3.5. *Human Resources involved during the discovery stage*

The Process Manager chooses the digital model of a system from the last DELTA-type project (less complex system *a priori*) as a standard to allow suppliers to establish their quotation. Then he writes down specifications uniquely containing functional specifications: the elements of the main part to develop, the target process, the actions required of the supplier and the target schedule. These specifications are not approved by the Purchasing Expert; and no other correction method is carried out.

This phase starts by making contact with MixMat, firstly by telephone. Then MixMat agrees to sign the confidentiality agreement put forward by DELCAR. To respond to the consultation on the DELTA project, MixMat selects a commercial counterpart who would become the unique point of contact with DELCAR on this project. He is the person who puts together the specifications and issues the first quotation.

3.2.1.1.4. Performance of the relationship

During this discovery stage, a relationship is developed between the two companies – the first goal is therefore complete.

3.2.2. *Exploring the possibilities*

September 2012

Mario issues his specifications to four potential suppliers and also provides a digital definition (DFN) of the DELTA system: This is a consultation digital mock-up "*(which) is a digital definition where the mold already exists and serves only to rank each mold-maker(supplier) relative to the other. The real DFN that will be used for the requested mold will be different, but the principle is the same*". He also travels to two of their offices to present the project to them. After these exchanges, each organization sends Mario a quotation.

October 2012

A new resource is available in the Innovation-Purchasing department. The Innovation-Purchaser, Asclepius, arrives and is present for two days a week for the next few months. He comes to support the Delta project team and helps Mario select the suppliers.

Mario therefore provides him with the rest of the elements in his possession: the initial functional specifications and the proposals that he had since received from suppliers. He then highlights MixMat and a second supplier as they "*seemed to have the best skills for this application*". He then informs him that they have to decide on a supplier within a month. This is so they can involve the supplier in the design of the definitive DFN in order to predict technical difficulties regarding the implementation and the stability of elements inserted in the mold.

After this handover, the Innovation-Purchasing Director asks the Project Manager that Eddy, the Expert Purchaser, specialist of series mold purchasing, is also involved in selecting the suppliers. Mario responds, saying that he "*raised the subject with Eddy and that he thought the mold makers he visited in Portugal were not necessarily adapted to what we want to do*".

Innovation-Purchaser and Researcher – the observant participation, a method inspired by ethnography

The DELTA project has been studied from a more current ethnographical approach among sociologists than among management researchers, using observant participation. It required the researcher to join the Innovation-Purchasing department for a few months and to participate in their activities as a member of the team. The author is therefore Asclepius.

In the organization observed, the observant participant from his side is identified as both a firm actor and a researcher in strategic management. Throughout his presence, his priority was to participate in the organization's activities, the observation being made on the basis of what he could experience and observe (Figure 3.6).

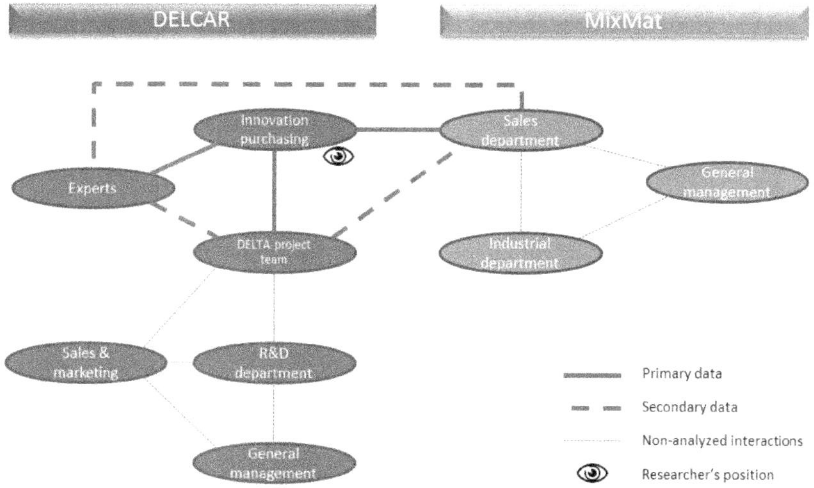

Figure 3.6. *Interactions observed between the client–supplier partnership. For a color version of this figure, see www.iste.co.uk/servajean/dynamics.zip*

This methodological approach allowed him to be at "the forefront" of the action, also limiting rationalization bias *a posteriori* and giving him access to a large variety and amount of data on the partnership's operational progress (Figure 3.7).

Primary data	Description
Observant participation	Participation in cases with the role of Innovation Purchaser Electronic and paper research notebooks which compile all the notes taken throughout the observation and emails exchanged – 550 pages for 2 days a week for 20 months in the field Presentations and documents produced thanks to the researcher's participation Contracts Further interviews

Secondary data	Description
Project documentation	Presentations made by the project team for passing phases in the Innovation Project's management processes
Internal data of the quality system	Descriptions and documentation on the Purchasing processes and the Innovation Project's management process
External data	Company websites Newspaper articles Informal interviews with members of the business' ecosystem
Increase in confidentiality	Submission of the narrative to the Purchasing and Innovation Director to increase the confidentiality of the data

Figure 3.7. *Data collected for this ethnographically based approach*

By being "a worker among the workers", the researcher could have access to data which they would not normally be able to gain access to. By participating in operational work, they make visible situations considered "too trivial" to be considered by existing theories [JOU 08, pp. 70–71]; they can then build daily operational management theories.

This participation in the life of the business also provides observation opportunities linked to the possibility to manipulate several minor events. In the case of the DELTA project, the researcher frequently requested informal assessment of the quality of the relationship or of the importance of the supplier chosen for DELCAR – assessments which would not have taken place outside this research setting.

Asclepius then meets the experts that were already involved in the project. The Material Experts explains to him the different technical constraints relevant to their specialisms. The Purchasing Expert puts forward to him the idea of involving other potential suppliers. The expert says to him that: "*We have some companies in Portugal who know how to design then create in series, if they are given the work and I have already worked with them. They have 2 or 3 really good teams and who have potential. If they want a helping hand, I can get in touch with them but it must be legit. What they present to one, they must present to another. These are expensive tools: €250-300k... they must have product specifications and specify what will make the mold. We will have a feasibility feedback. The specifications must include: material, cycle and output*".

In the first coordination meeting organized between Asclepius, Mario and the Project Manager, the latter starts with an introduction that outlines "*the most important thing is that DELCAR does not take responsibility for the mold if there is a problem. Internally, there will be a quality control document which makes it clear that the supplier commits to the quality of the mold and that it works*".

He does not wish to relive the experience they went through with a supplier involved in another innovation project: "*by involving the supplier we give them more responsibility. The question is, who pays if we need to remake the mold to guarantee the quality of the part? Who pays for the modifications? We cannot start this project like we did with the other*".

For Mario, his job to carry out with the supplier is important in order to find solutions to complexities and take into account the novelty of the DELTA system: "*We need to design the part according to the mold*". He is also conscious that he cannot ask the suppliers "*to quantify the true project*": the true project being the system designed definitively. As Manufacturing Process Manager, his expectations are not only technical (equipment and specific know-how) but also collaborative: the supplier has to be able to contribute to "*assisting with the concept, which does not have a final definition*" and "*we must be able to work together, come up with ideas together... be seamless in the manufacturing of the part*".

A meeting is then organized between Mario and Asclepius, with the Purchasing Expert and the Material Expert. In this meeting, it becomes apparent that there is no way to compare the supplier's responses how they are at the time. They do not cover the same perimeters, and the metal used to make the molds differs too much. The prices vary from single to double.

Eddy also started to consult three other suppliers and understand their interest in such a project. He suggests rewriting the specifications entirely so that the suppliers can respond with the same perimeter. It is important for him that "*his*" suppliers are treated properly as he wants to maintain a good relationship with them. "*In exchange*" for the presentation of "*very good suppliers*" to the R&D department, "*the R&D team needs to be fair, and present exactly the same thing to every mold maker*".

"*Over there in Innovation* "says Eddy" *it is an imaginary world and here (in Purchasing), it is real life. I don't want to ruin all the fieldwork from the past 15 years. It's the same thing in Purchasing. They are not taking everything into account, they risk making nothing of worth and me, I don't want to ruin everything, firstly because I have not finished my professional career and secondly, because they trust me. By initiating exchanges, by giving a helping hand in one area and then in another, we built something in line with what they requested, which they can take out there, even freely*".

Then, Eddy provides the Innovation-Purchaser with technical specifications for the manufacturing of the series molds, instructing him not to pass it on[1]. According to him, the selection of the suppliers, based on precise specifications, is made "*first on the ideas and then the price. But also, on the question of knowing if they will stick to the project or not, and to DELCAR. We do not have a reputation where we fulfil our developments. We promise a lot then afterwards we have to manage things alone. More specifically, this has happened for the past few years. It has started to make its way round (within the sector) where everyone knows. It's not that we are bad but that we are short staffed, so suppliers get their c#!§%µ* back.*

DELCAR suffers the consequences: loss of time, it's the toolmakers who escalate the problems, for example, that we have not paid for all the work and they are led to do more work – the argument is that they have already done their job – after a while, they say stop. There are a lot of good suppliers who refuse to work with DELCAR in plastics processing and also in tools. In plastic processing, there are two of them who have said no, but the majority have not said no outright, they just don't respond to our RFQ[2] *(consultations) anymore. The selection also has a personal angle, so I warn them about competitors on the project to please them, so they make a nice offer*".

A draft copy of the specifications is drawn up by the Innovation-Purchaser by taking elements of the project innovation specifications in a framework based on the specifications of the series mold. It is defined and completed by technical data (tonnage capacity, surface aspect of the finished part, thermal regulation range, etc.) gleaned at will from interviews carried out with different project participants, members of the project team and experts.

---

1 Such specifications are in fact considered by the Expert Purchaser as a personal tool to protect; it is a part of what makes him an expert and assumes his fragile position, within the firm.

2 RFQ = Request for Quotation: a standard business process whose purpose is to invite suppliers into a bidding process to bid on specific products or services. RFQ generally means the same thing as a call for bids (CfB) and an invitation for bids (IfB). To receive correct quotes, RfQs often include the specifications of the items/services to make sure all the suppliers are bidding on the same item/service. Logically, the more detailed the specifications, the more accurate the quote will be and comparable to the other suppliers: https://en.wikipedia.org/wiki/Request-for-quotation.

Through a presentation used in previous projects by the Director of Innovation-Purchasing, the specifications are reconstructed. The evaluation criteria for the supplier are also added:

– ability to satisfy our needs;

– performance level and quality of means;

– creativity (modification proposals, new technical solutions, optimization of the process etc.);

– desire to cooperate;

– complete costs;

– responsiveness;

– reliability;

– knowledge of implementing material α;

– sensitivity to rheology;

– effort put into the response to RFQ.

A decomposition table on the supplier's costs is added using a model for the purchasing of series mold[3]. Elements of this specification are still to be completed (mechanisms, injection nozzle type, methods for the prehension of the inserted elements, etc.).

November 2012, after the start of the following week, the Director of Innovation-Purchasing announces that consulting suppliers for the DELTA project are to be left to the Innovation-Purchaser for the Purchasing/Commercial side.

The provisional specifications serve a basis for a new meeting between the Innovation-Purchaser, the Process Manager, the Expert Purchaser and the Director of Innovation-Purchasing: "*the objective of the meeting is to clarify things and that everyone fully understands the requirements*".

3 The costs decomposition table makes it possible to compare each supplier on each entry, in order to negotiate with each of them to achieve the lowest costs. It is a common purchasing method.

The requirements for those involved in the innovation project are:

– *"development of a mold allowing us to make at least 100 parts*;

– *cycle-time target = 60 seconds;*

– *sequential injection;*

– *taking into account the rheological study which has already been carried out;*

– *co-development for the design of zones posing technical difficulties;*

– *supplier equipped with a Z robot to place inserts into the mold;*

– *supplier's competency working with the material α family on trim pieces;*

– *testing the produced parts and the carrying out of capabilities;*

– *development of control means if necessary*".

A guarantee of five years or of 2000 parts is requested specifically by the Purchasing Expert, and this is "*to be sure that we have a high quality mold that can perform its abilities in series*". These elements, of which some are new (minimum tonnage capacity, tool guarantee, quality standard, selection criteria, cost allocation table) are integrated into the new specification.

Furthermore, with regard to the status of the suppliers' responses, all are considered by Mario to have correctly answered the request "*they have responded to what we have sent and have taken the following approach. Do as you wish*". Following this meeting, the new specification is sent to the consulted suppliers by the Innovation-Purchaser – the digital definition used for costing the first remains unchanged.

A week later, Mercury, MixMat's Business Manager, revises his quote having contacted Mario by phone. He indicates to him that the changes requested require him to only modify the metal of the tool. So MixMat only revised its quote on the lines impacted by this change.

A new metal category is proposed: V metal (a category considered inappropriate by the Purchasing Expert). Copied into this email are the Industrial Director, the Commercial Director and the Marketing Manager of MixMat. The other elements do not appear to have been taken into account; the cost distribution table remains the one initially proposed by MixMat.

The Innovation-Purchaser then contacts Mercury to confirm if he took into account the desired process in the costing, along with the Z robot. This indicated to him that if he has the facilities and knowledge for the desired process, he has only quantified the bare mold without particularly taking into account this process.

On this occasion, he informs the Innovation-Purchaser that Mario will come the next day to visit the firm and see MixMat's facilities. Also, the Innovation-Purchaser indicates to Mario and Mercury that they should on this occasion rework the comprehensive requirements, thus going beyond the bare mold; and that following this, the costing shall be reviewed. Meanwhile, the three suppliers outlined by the Expert Purchaser have declined the RFQ.

The following week, a meeting is organized upon the DELTA Project Manager's request, in order to provide an update on the selection of suppliers. The manager of the Material Experts wishes to be present. In addition to these people, the Innovation-Purchaser and the two Process Managers are present.

The Project Manager begins the meeting by announcing that "*the context of the project is tense with VROUM as the relationship has deteriorated at the higher management level due to industrial policy issues*". He creates an evaluation table showing the offers of suppliers who have responded: the positive and negative points and a ranking for each actor involved.

Mario offers his opinion on MixMat following the visit that he carried out at their premises. MixMat has given "*a very good impression. They have old machines, but they work. They work a lot with (a renowned test center) as they don't have their own test center*".

The Experts Manager announces that a collaborative project involving DELCAR and MixMat, along with other companies in the automotive sector, has just been submitted for public funding.

In front of another supplier whose quote is 30% higher, but whose quality is guaranteed, MixMat appears to be a challenger. Their tools and their geographical proximity are positive points. For Luigi, assistant to the Process Manager, who also visited MixMat, "*they pinpointed our problems*".

All the members involved in the meeting rank MixMat in first place for their response. For the Project Manager "*MixMat have an R&D strategy in line with ours. They work in the technical way we desire. Is that the reason for the price? The situation has reversed, I thought we would have gone with the other supplier*".

They then decide to ask the two finalists for a complete quote again. Mario sends the summary table to the missing experts as well as to the Material Expert from the other division of the DELAUTO group. Then he passes it on to Asclepius at his request; to be forwarded to the Innovation-Purchasing Director and the Expert Purchaser.

During the internal debriefing of Innovation-Purchasing, the Innovation-Purchasing Director indicates to the Innovation-Purchaser how he would work on the project. The Process Manager has to indicate to the suppliers that he is the point of contact for any technical question, while the Innovation-Purchaser is the commercial one. The Director of Innovation-Purchasing states:

"*As a buyer, the job is to prime the pump and stick to MixMat's quote budget even if they have forgotten about it for now. You have to ask them to fill in the cost breakdown table and inform them that otherwise it will be difficult to go further. The language we use is important because you have to be firm without closing the door. You never know if after that there is only this supplier to choose from. It's a game*".

The expert from the other division of the DELAUTO group suggests that the consultation be reopened to the initial category of V metal (the hardest metal): "*it was a mega b@ = &%!... to make a Y metal mold prototype to approve the cycle time because it will never mold at a quick enough rate, by heating up as it should. So, we can only simulate it mechanically but not for the cycle time*".

The Expert Purchaser and one of the Material Experts shares the same opinion on the impossibility of approving the project's cycle time in the project. Also, the complete costing will be requested from the third supplier who only manufactures metal Y molds (the least hard). All three suppliers are required to review their offer and complete the cost breakdown table for this project. This is no problem for MixMat's Business Manager.

A few days later, the DELTA project can be cancelled at any time.

The executives of DELCAR and VROUM do not agree on the scope of the project. DELCAR wants to widen the scope when VROUM wants to restrict it to thc single part, which would negatively affect the profitability of the project for DELCAR. If VROUM will not change their mind, it is better for DELCAR to continue with traditional solutions rather than embark on a new and risky process with a higher cost. Pending a decision, the payment of the budget is put on hold by VROUM. At the same time, DELCAR's policy on this type of innovation has changed: lightness now takes precedence over the issues of appearance – unlike the situation at the launch of the project, a few months ago.

Dionysus, the Material Expert specialized in mold design, is consulted by Asclepius. He believes that MixMat is not in a position to handle DELCAR's requests due to the inadequacy of its Technical department. He also thinks that some technical answers in their response do not show that they can develop the requested tool. He relays the rumor that MixMat will no longer work with DELCAR's main competitor. Finally, he considers that the cost of calling the test center has been underestimated. On the other hand, he recommends using V metal to develop the tool.

In order to reduce doubts about MixMat's financial stability, the Innovation-Purchasing Director asks that the suppliers in the lists fill out a supplier evaluation form and an evaluation form of their R&D. These documents, resulting from DELCAR's Purchasing quality procedure, are retrieved on the Intranet by the Innovation-Purchaser. Only the first is sent out, with the R&D evaluation form not being suitable for this type of work.

MixMat's Business Manager is asked to confirm that the costs related to the test center have been taken into account; this is done by return mail detailing these costs "*in agreement with Mario*". With regard to firm accounts, MixMat requested in the supplier evaluation form that their Business Manager specify by telephone that "*the firm has completely changed, in that our accounts have been cleared*".

At the beginning of December, during a working meeting on the comparison of the cost breakdown of the offers between the Innovation-Purchaser and the Process Manager, the latter declares himself bothered by the rumor of MixMat's de-listing by DELCAR's competitor. In the meantime, he also receives an email from an Expert from the other DELAUTO division who writes to him saying that he should choose the third supplier because he is already working with them. He also tells him that it is necessary to choose a solution with the Y metal because it will not be possible to validate the cycle time even with V metal.

At this point, for Mario, the choice of supplier cannot be made until there is an answer to the question "*what is expected of the mold?*".

A few days later, VROUM agrees to participate in this phase of the DELTA project.

The DELTA project continues; but the decisions on the scope of the project during the industrialization phase have not been decided. A meeting to advance on the exact need (prototyping the product and the process, with the V metal or prototyping the product with the Y metal) and on the selection of the supplier (according to the predefined criteria and presented in the specifications) is organized by Innovation-Purchasing; however, it is cancelled due to a lack of participants.

The Innovation-Purchaser then informally evaluates the position of key decision-makers. MixMat is recommended "*wholeheartedly*" to the Material Experts Manager by a firm that uses them for all its molds. The Process Manager "*leans more towards MixMat because the defending supplier takes a lot of time to answer and (it) seems complicated to manage. (I had) MixMat on the phone who to him seemed motivated*". He will visit the third supplier the next day with a material expert for another project.

All the data requested from the suppliers arrives in the meantime. The different cost items can be compared:

From: Asclepius

"The hourly rates are similar, as are the prices of the different components (a few variations but these are minor proportionally). Differences in price come above all from the supplier's evaluation of how long it will take to complete the project".

Following the third supplier's visit, this one convinced Process Managers that they had to work with them for another project. So, they are also considering entrusting them with the DELTA project; this suppler is also the cheapest of the three. For this reason, they organize a meeting on the innovation floor, without Purchasing present. During this meeting, this supplier is unanimously considered "*as a good choice; if this was not a question of (inserts), it would be them without doubt, but there is this risk*". These elements are not passed to the Innovation-Purchasing department before the selection meeting that will take place the following week.

Organized by the Innovation-Purchasing department, this meeting brings together the Project Manager, the two Process Managers and two Material Experts, the Material Experts Manager and the Expert Purchaser.

It is led by the Innovation-Purchaser based on a projected presentation that presents the elements collected to choose (1) the need and (2) the supplier that best meets this need. Among them all, there is a consensus that the cycle-time is not so important. The possibility of requesting a pre-study on the main issues related to the choice of metal tool is then suggested: which class of metal for the mold is likely to wear down on the inserts early on?

MixMat's Business Manager proposes that a first study be done on this point before designing the mold as a whole. The Project Manager offers to pay the suppliers to do this. For Mario, MixMat will not be "*sensitive to this type of proposal*". Their interest is to have the whole project and their expertise is to look for "*super innovative technical solutions*".

At the end of the meeting, only MixMat and the third supplier remains in the running; the first supplier is largely the most expensive of the three, the least responsive and the one with the least suitable response. It is decided to meet the two finalist suppliers within one month. They will be asked to perform a mold design study for the sub-group posing problems. This study will be financed by DELCAR in connection with the DELTA project-coordination team.

The two suppliers are asked to provide a quote on the current system DFN. The meeting report, drawn up by the Innovation-Purchaser, also specifies that the selection of the type of metal and the supplier will be done "*after analysis of the costing and the design. The quality of the collaboration will also be taken into account (there are still doubts about the (other supplier's) ability to evaluate the necessary co-development efforts)*".

3.2.2.1. *Analysis: exploring the collaboration – first exchanges and questions*

This period is characterized by the engagement of discussions, the first negotiations and the first phase of interactions in the relationship. The first substantive exchanges between organizations followed the signing of a confidentiality agreement.

The first evaluation of the needs and resources of the other was done through the transmission by DELCAR of the first version of its RFQs (the specifications and the digital definition of the standard part), then by the answer provided by MixMat (the quote and its general conditions of sale).

Other intercorporate exchanges included visits to MixMat's premises and the exchange of updated versions of the specifications and quotations. This consultation was done while consulting other potential suppliers at the same time.

3.2.2.1.1. The atmosphere of the relationship

This episode is marked by increasing knowledge of each other's needs and resources. It is also marked by DELCAR's increased trust in MixMat to meet its needs, in absolute terms, but also relative to some of its competitors (Figure 3.8).

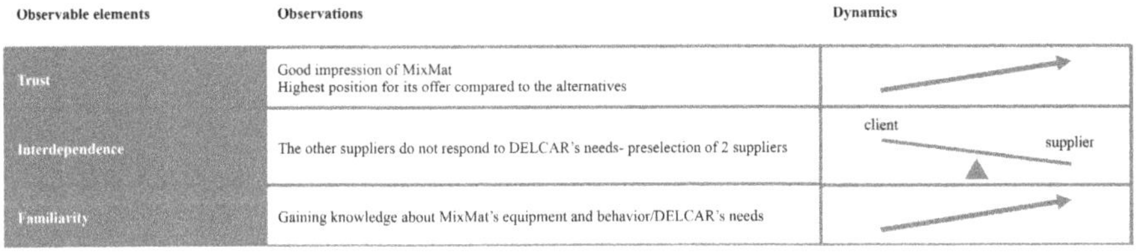

Figure 3.8. *Atmosphere of the relationship during the first exchanges*

At the beginning of this phase, DELCAR carries out consultations with other companies that should have been able to perform the same tasks as MixMat. As DELCAR exchanged with the various potential suppliers, it turns out that few companies meet DELCAR's needs.

After analyzing the responses made by MixMat, it is considered to be one of the few companies capable and willing to work with DELCAR on the subject.

The visit to MixMat's premises[4] and then telephone interactions reinforce the trust of the project's technical players in MixMat to respond to the project's problems, in absolute terms and in relation to other suppliers.

"*Following our visit, they made a very good impression. They have old machines, but they work*" says the Process Manager at the start of the phase, and two weeks later: "*I am leaning more towards MixMat because the defending supplier takes a lot of time to answer and (it) seems complicated to manage. (I had) MixMat on the phone who seemed motivated*".

When rumors arise about the lack of MixMat's financial stability, checks are carried out by using tools from the Procurement process. The results of these checks then reinforce DELCAR's trust in MixMat.

3.2.2.1.2. Maturity of the innovation project

Uncertainties about the technical solutions that will make the piece work are still strong. This leads to them examining the supplier's technical evaluation further and reinforcing the client's need to integrate it as soon as possible; the finalization of the main part's design depends on the definition of its manufacturing method.

Uncertainties as to the commercial exploitation phase of the DELTA project increased during this period. The lack of definition regarding the project target (a product prototype or a product and process prototype) balances the choice of supplier.

At the end of the phase, there are two suppliers considered for the rest of the project: MixMat is better suited to the second possible target, and the other to the first.

3.2.2.1.3. Management of the relationship

The first exchanges between MixMat and DELCAR are followed by a new resource at DELCAR, the Innovation-Purchaser (the author), in the Innovation-Purchasing department. He becomes DELCAR's commercial representative for this relationship with MixMat (Figure 3.9).

---

4 For another supplier, on the same project, the visit of the Process Manager led to an equivalent increase in his positive impression of the supplier as to his ability to meet DELCAR's needs.

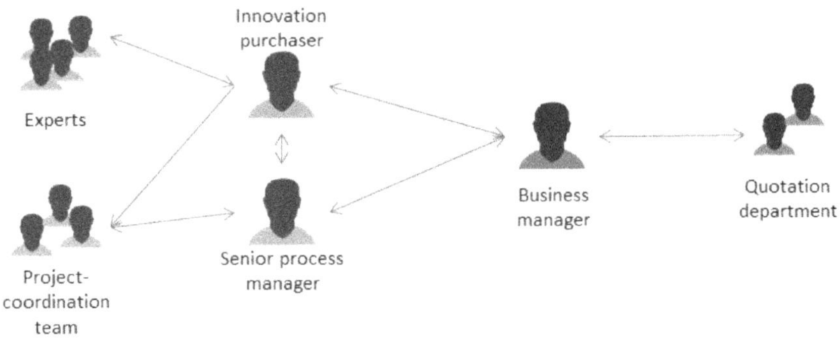

Figure 3.9. *Human resources involved in the first exchanges*

The Innovation-Purchaser returns to the initial specifications and consults different experts to specify/ensure that: the functional specifications are expanded on, and the technical specifications are added. The editing of these specifications is accompanied by the development of a spreadsheet by different DELCAR counterparts (Innovation-Purchasing, Experts and Project team) to allow them to compare different responses.

These new specifications, including functional, technical and commercial elements, are sent to MixMat in order to obtain an updated costing. The Business Manager responds to it with a new quotation. Throughout this phase, he also responds to the different requests from the two DELCAR counterparts, on the technical as well as the commercial points. To put together the quote sent to DELCAR, MixMat's Business Manager uses the Quotation department of its firm which belongs to the same Direction as him.

In order to confirm MixMat's financial status and the contents of their Technical department (and therefore its ability to respond to technical problems), a framework from DELCAR's Supplier Management System is used and submitted in addition to the project-specific elements.

The evaluation of MixMat's ability to respond to the functional needs of the DELTA project is carried out through a visit by the two Process Managers to their premises. The evaluation of MixMat's responses, and its comparison with the other companies' responses, is carried out by different participants at DELCAR.

3.2.2.1.4. Performance of the relationship

During this phase, the efficiency of the relationship can be evaluated through the convergence between different requests carried out by DELCAR (the different specification versions and the requests to conform to the response formats) and the responses given by MixMat. This convergence is not cemented at the moment: MixMat, like the other suppliers consulted, only partially meets the specifications and does not meet all the requested formats. However, it meets them relatively more than others.

3.2.3. *A "false start"*

The last week before Christmas

Mario and Asclepius have to call the two suppliers to ask them to carry out a preliminary study on the choice of metal for the mold in regard to inserts of the DELTA system. This is also to arrange a meeting at DELCAR's office. Together, they prepare a presentation on the pre-study that Mario has to send out the day after.

At MixMat, Mercury is not reachable. Mario and Asclepius do not want to waste time so they send this new information to him by email:

From: Asclepius

Good afternoon sir,

I was not able to contact you this afternoon by telephone, therefore I write to you to inform you that we have selected MixMat to carry out work on the DELTA project. We would like to discuss the following with you:

– we will send you the final DFN for a definitive costing by the 3rd week of January,

– we would like MixMat to carry out for DELCAR a pre-study for DELCAR on technical point X (the first step you suggested) in the same timeframe, which will be financed by DELCAR.

If you could please contact Mario so that he can explain to you how we can move forward on the subjects discussed above and to arrange a meeting at our office in the 1st or 2nd week of January?

I look forward to your reply.

The morning after

Mario sends both suppliers an email including this one to MixMat:

> **From: Mario**
>
> Hi Mr Mercury,
>
> Following on from Mr Asclepius' email, please find attached the information, if you agree with the approach. You can contact me if you would like to discuss this with me in person.
>
> I will send you other digital information separately by CD.
>
> Kind regards,

Following the two emails to MixMat, the Innovation-Purchasing Director sends an email to MixMat's Business Manager to confirm that the two previous emails are neither acceptances of the quotes nor orders[5]. At the same time, he explains to Mario and Asclepius the commercial risks linked to this type of email.

> **From: Orion**
>
> Mr Mercury,
>
> The email below from Mr ASCLEPIUS following the one sent today by Mr MARIO (request for costing regarding a pre-study with a confidential document attached), have been sent only as a request for quotation for which you have been selected.
>
> The two emails are not in any way the go ahead for this request or any on behalf of DELCAR. However, they do not affect the elements and scheduled meeting that will allow you to better understand our needs to finalize your offer.
>
> The two previous emails could be confusing, so I wanted to clarify this important point.
>
> Kind regards

5 The quote is not in itself a contract, but a unilateral commitment that becomes a contract only when the quote has been accepted by the person to whom it is delivered. Also, these exchanges could be regarded as commitments on the part of DELCAR vis-à-vis MixMat.

Following this email, Mercury leaves a message on Mario's answering machine to tell him that, under these conditions, he does not wish to work with DELCAR. He is not ready to move forward and it is not suitable for them.

Later in the day

Mercury emailed Mario again to say that he wants to reconsider the question of working with DELCAR. He acknowledges receipt by email of the new specifications Mario has sent and announces that he will propose meeting dates after the holidays.

3.2.3.1. *Analysis: a hiccup in the relationship*

There are only two suppliers left.

The goal of what will be asked of them balances between two targets. For the client, choosing this target results in choosing one or the other.

In order to start the investigation, DELCAR chooses to ask them to carry out a pre-study which would allow them to decide between them while moving forward with the project.

Two consecutive emails are sent to MixMat by the Innovation-Purchaser and the Process Manager. The terms of these emails could have indicated DELCAR's commitment to MixMat for the DELTA project – and this is how the Business Manager of MixMat perceived it as well as the Innovation-Purchasing Director of DELCAR. Also, as soon as the latter is aware of this, he sends an email indicating that this is not the case.

In 24 hours, MixMat's Business Manager received three emails: the first two seemed to suggest to him that his firm has been selected – that his quote had been accepted – then the third indicates that this was not the case and that there was no commitment. Due to these ups and downs, he wishes to cease the relationship with DELCAR on the innovation project. Also, by voicemail, he indicates to DELCARs Process Manager that he wants to end all discussions.

Several hours later, he changes his mind. All this, 48 hours before the Christmas holidays.

The tension that arises from this situation will ease over the Christmas period, when the two companies are closed.

This situation marks a hiccup in the relationship whereby DELCAR feels there is a possibility that they can "lose" this supplier, who could be capable of responding to their needs.

The atmosphere of the relationship has changed: trust and interdependence marked a dip in the relationship.

3.2.4. *Exploring a potential collaboration*

January 2013, return from break

VROUM wants to see its budget for the DELTA project decreased.

The Innovation-Purchaser is, for the first time, the recipient of the minutes of the project-team's weekly meeting. It announces that the DFN for the final costing will have to be done by the end of the month

Mercury, MixMat's Business Manager and Mario, the Process Manager of the DELTA project, speak on the phone. MixMat, like the other supplier, does not want to do a pre-study, even though there is a need to develop a first tool, without doing the mold: "*We want to know where we are setting foot*". Mario responds to him that "*someone from Purchasing will call you for the strategy to follow*".

The Innovation-Purchaser then makes contact with Mercury by phone. The conversation starts by going back to the last exchanges before the holidays. MixMat thought they had been selected. Then "*we received an email from the Innovation-Purchasing Director that nothing had been done yet*". A long explanation follows:

"*It's been years since I've done development internally and I have carried out developments with many different types of companies, from high-tech to other automotive companies. I don't want my ideas to be stolen. I know that I can carry out your tests and carry out the equipment. I don't want to be squeezed on the price. With Mario, I have no problem technically. We agreed on prices at the start. I will be consistent, and I do not want to take over: I want to work as a partners. On the other hand, we must be sure to have a good partnership behind us... it's ok for this first costing. I will send the technical solution costing to Mario this week. The idea is that this technical solution is replicable on a series level – as a prototype it's easy – the difficulty is integrating it... The idea of a partnership does not bother me – the moment we move to R&D, we have to work as partners*".

In order to carry out the costing for this work of producing a "pre-study tool" and carrying out the study itself, Mercury has to establish contact with Mario to evaluate the technical feasibility. In the morning, he sends him the plans explaining the technical principle and asks Mario to call him back to finalize the costing.

Following this telephone call, the Innovation-Purchaser joins Mario to discuss what is to come. Mario indicates to him that he is "*bothered by the non-choice of suppliers. I need to know if they are going in the right direction in terms of the DFN... And that is starting to be very tense*". He has no doubt about MixMat's ability to create the requested tool and carry out the tests.

Compared to the other supplier, he believes that, with MixMat, it will only be more difficult to "*play around with the product*", i.e. to proceed directly to a series of changes and tests on the mold and the part to produce. This can be useful, however, since DELCAR does not have the capacity to experiment – this opinion is shared with one of the Material Experts.

At this stage, it is envisaged that they will request the development of two different tools in order to produce the target part. The Innovation-Purchaser lays out two possible routes: (1) a tool which is more guaranteed to achieve the desired result and (2) another which will allow them to explore several leads and learn about this type of project but with less assurance with regard to the result.

The Process Manager indicates to them that it will be the "*technicians*" who will decide which route to take. The Innovation-Purchaser himself indicates that it must be discussed with the Project Manager and the Innovation-Purchasing Director, this being a strategic decision. This exchange, which also involved the Process Manager's assistant, leads them to question the consideration by MixMat of "*all the subject's complexities*".

The day after, during an informal conversation, the Project Manager says to the Innovation-Purchaser that, at this stage, it is not guaranteed that the DELTA project will move forward: "*it has no more than a 50% chance of succeeding*". The DELTA project has not been included in DELCAR's marketing plan. DELCAR's logistics facilities at that time do not allow them to guarantee the delivery for the target vehicle. The last price estimations do not allow them to put forward a price which is equivalent to traditional solutions. Finally, DELCAR's sales have reached the limit of VROUM's "supplier dependency regulations".[6]

For the Project Manager, this type of product will not to be competitive if they have been able to produce it: either they do not have the financial capacity to develop it, or their production capacity makes it possible to produce the classic solution, but this innovative solution "*would lead to the closure of factories*". This explains to them why the DELTA project is not included in the development programs plan. "*Also, as for the choice of supplier, I prefer to ensure conservative investment management and have a cash neutral operation by selecting the cheapest option as this will also allow us to limit the final cost of the innovative DELTA system*".

In mid-January, the other supplier meets DELTA's Project Team at DELCAR's office.

The same meeting is scheduled at the end of January for MixMat.

Before this, Mercury sends a quotation to Mario and Asclepius for the supplementary pre-study tool needed to carry out the test. This subject is no longer "*urgent*". Mario received a new email from Mercury, who "*will go with the ideas but who thinks that the technical difficulty requiring the development of a first test tool is impossible to replicate in series*".

The Innovation-Purchasing Director then asks Asclepius to put together a comparative table of the 2 suppliers' offers, leaving the price highlighted, as this has not yet been negotiated. He also requests that the specifications be included in order to ensure that the suppliers are 100% compliant.

6 A customer's dependence on a supplier may jeopardize its supplies. Also, large groups generally plan to limit the share of their purchases in a category to a supplier at a level around 30%.

The meeting with MixMat at DELCAR's office includes MixMat's Business Manager, Mercury, the DELTA Project Manager, the two Process Managers and their direct superior, two Project Engineers, the Material Expert specializing in mold design, Dionysus, and the Innovation-Purchaser.

Mercury starts by presenting MixMat, its history and its innovation and production abilities, the technical activity fields they work in, their current clients, their employees and their recent acquisitions. The first exchanges were with regards to questions surrounding the development of equipment and their thermoregulation to develop complex parts for the aeronautics and automotive industry.

Mercury then shows preparatory work he has done from the DFN sent to him. The discussion takes place around the different areas of difficulty. Mercury asks questions and asks about changes between this DFN and that which was sent to him initially. He makes some suggestions illustrated by projected images "*that I would like to put forward to ensure sufficient sealing: a bracket is provided which fits into the hinge – rather than a mold – before putting insert G into the mold. It is used to provide peripheral sealing. In addition, it facilitates the insertion of the part into the mold*".

Then follows a discussion between the engineers on the possibility of ensuring the industrial feasibility with this type of solution. The part in question is then physically presented to Mercury. This then shows how easy it is to position this part into the mold. Mario asks him if it is possible for the MixMat solution to meet the automobile cycle times. For Mercury, "*this bracket is mechanized with a compartment to put it in automatic. I have more doubts*" (even though in the next hour he will say "*I have doubts about the industrialization series of the bracket put on insert G*"*)*. Mercury then presents a diagram on paper representing the complete equipment.

Exchanges are also made around the room posing difficulties as well as a prototype system prefiguring the one that will be developed. Around drawings made during the session by Luigi, the discussions are centered around the target process, mainly between the DELCAR engineers. A Project Engineer then shows plans "*summarizing point G's difficulties*" then the rest of the DFN to highlight the other features of the system.

A discussion on the other main technical concerns is supported by whiteboard drawings – DELCAR engineers make many digressions between them about the feasibility of a particular option. For each point of difficulty raised, Mercury proposes a solution or validates the feasibility of a solution envisaged by DELCAR or MixMat, with the help of plans on the whiteboard or on his computer. Later, a Project Engineer indicates having learned during this meeting how, for a prototype mold, adjustments are made in steps of a few millimeters to then size the dimensions to achieve the mold-series.

The Project Manager, before leaving the meeting, asks Mercury what MixMat's offer would be. To this he responds: "*I wanted to discuss with you approving the offer in line with the first costing we did, there is more to it than we had predicted. Given that we have many things to refine, we are not in the same situation, without going half the distance. It will not be sc@ µ * - & up...*".

Several solutions has to be envisaged by MixMat to answer the difficulty of the G-insert, without resorting to the MixMat Research & Innovation team. "*The idea is to use the fact that we are a small team to make the best use of everyone's knowledge... if I do not have a solution... I will not propose one*". Mercury asks for additional technical information on the dimensions. They will be sent to him by email.

Planning issues are also addressed. Mercury says he will send a new costing the following week. For Mario, they have to start working together within one month, i.e. by the end of February. The Innovation-Purchaser requests that a costing be put together using the costing matrix that he sent. For Mercury: "*This does not pose a problem and now we know where we are. Next week, I will talk with the quotation department. Firstly, we will evaluate the mechanical solution to be provided. Secondly, we will assess if we can create (the other) version even if it is more complicated. After all, we have very experienced guys.*

– "*You must provide a costing for the second point*", the Project Engineer informs him

– "*From there we can launch a prototype*", Mercury replies to him. "*We can think of an automated solution for the future*".

– "*At DELCAR we try to be the closest to the series. We're at the stage where we're doing prototypes. We want to validate the process a little bit*".

The meeting is then finished after a final discussion on questions surrounding the main part.

Following this meeting, the Innovation-Purchasing department organizes a meeting with the DELTA Project Manager and Mario in order to "*analyze the 'supplier' specifications for the DELTA project. As we move forward with the project and the meetings, the needs are evolving... We need to stop our enquiries . This will allow us to choose between the other supplier and MixMat in a level-headed manner*".

To prepare for this meeting, the Innovation-Purchaser creates a table which recaps all the specification points and certain technical points highlighted within meetings with suppliers. The latest DFN, which is not definitive, has been sent by Mario to MixMat for a costing.

For the Project Manager, the study has to be launched by the selected supplier within two weeks to prevent any delays, and in three weeks there will be a monthly innovation committee meeting, where it will be decided if the DELTA project will be able to pass to the next phase. He recalls the importance of the quote and the scoping of the supplier.

"*I don't want to go through the same thing as we did with supplier K*", says the Project Manager... "*we launched a mold without specifications and without acceptance criteria... I do not want to have to pay them back like with the previous supplier where we had to make several extensions – it happened because we did not have a purchaser with us. As for the question of who we choose, we will have to make other prototype molds because the style will have changed. The problem is that there is a product with style and a crash test – it's watchmaking*".

The specifications are completed then reviewed a second time between the Innovation-Purchasing Director and the Innovation-Purchaser. Certain terms are changed: the Director changes the words "*develop*" to "*supply*" and "*will*" to "*shall*" so that "*legally the obligations are more on the supplier's side*".

These specifications are sent by email to the Process Manager and the Project Manager, so they can approve them. They will use them as the definitive specifications. As stated in this email, The Innovation-Purchaser consults with one of the Material Experts on the specifications. The Expert

Purchaser specializing in molds will no longer respond to any request – his professional situation has evolved.

Not all the comments made by this Material Expert are taken into account by Mario:

> **From: Mario**
>
> To: Asclepius
>
> I don't agree with everything. Certain things have already been taken into account, others are without a subject. Thank you for sticking to what we came up with together and please do not share this with everyone, this will not move us forward. I went over the objectives and schedule the other day.

A new costing is sent by MixMat 10 days after the meeting, and after numerous telephone conversations between Mercury and the Process Managers. The cost breakdown is done following MixMat's format. The Innovation-Purchaser therefore asks for Mercury to resend this breakdown adhering to DELCAR's format.

During an informal conversation between the Innovation-Purchaser and the DELTA Project Manager, the latter makes a point to him on the schedule: VROUM wishes to reschedule some of the tests, which will delay the supplier's delivery schedule by a month (to July instead of June). On the same day, the Material Experts team indicates to the Innovation-Purchaser their doubts surrounding MixMat's financial stability and their satisfaction with its involvement in ongoing collaborative R&D projects. Also, on the same day, the Process Manager meets, by accident, MixMat's Business Manager at a trade fair.

Several days later, at another fair, the Innovation-Purchaser meets a cluster coordinator who tells him he had met a person from MixMat: for him, it is very complicated working with DELCAR; it would be almost impossible if there were not collaborative projects in progress where the process is simpler.

At the end of the week, the two Process Managers indicate to the Innovation-Purchaser that MixMat is the most serious of the two suppliers in the running: MixMat "*thinks more about the series and therefore could be more appropriate for the project when the other is an experimentalist, a real prototypist who seeks by all means to make prototypes without thinking about the series*".

Furthermore, this other supplier does not respond to the request for a cost breakdown on the DELTA project. The Process Manager's assistant believes that, following his participation in a conference given by MixMat, "*if MixMat does not know how to do something, it has many partnerships and knows to call on them if necessary*".

It is at this point when MixMat sends their cost breakdown adhering to DELCAR's format. This is then compared with the other data previously gathered on the project and evaluated by the Innovation-Purchaser:

From: Asclepius

The differences in cost come to approximately 600 hours of further work and a much higher category of metal (which is twice as expensive) than the initial metal.

It is then decided that Innovation-Purchasing will start negotiations with MixMat, without completely eliminating the other supplier. Dionysus, the Material Expert specialized in mold design, warns the Innovation-Purchaser that one of his contacts in the industry told him that MixMat has a tendency to charge for any changes[7]. He therefore advises that DELCAR shall pay for the development until the delivery of the first compliant parts – this includes any modifications that will only be possible once the first tests have been completed.

Mid-February 2013

The Innovation-Purchaser prepares the negotiation with MixMat following recommendations from the Innovation-Purchasing Director: for him, it is important to prepare it by determining the level of importance and

7 In this way, a "classic" contract logic in controlled expenditures, which Midler has described as having a negative effect: the supplier offers a low theoretical first price, knowing that it will then be able to make money on the modifications that will not fail to occur during the project [MID 00, p. 33].

the ability to negotiate different points of the contract, for the client and the supplier; the point of the negotiation is to slowly pinpoint areas of difficulty, without reaching breaking point "*we do not want to ask too much of them and smother them*".

For Asclepius, the only point of difficulty identified for MixMat is the fact they are being paid a flat fee even though the DFN is still not definitive.[8] He also wonders about the possibility, outside of the negotiation, to make the package: the rheological study could push the boundaries. The simulation of the flow α material in the mold to form the main part could indeed lead to adding mass or elements to the tool.

In the meantime, DELCAR's practice for staggered payment of the tools is retrieved.

The following week

DELCAR receives the quotation from the other supplier. There are certain imprecisions, specifically related to the logistical questions, to which the Project Manager says, "*this highlights even more that we should choose MixMat*". For him, in the negotiations, the delivery date is the most important point, even more important than the price: "*for MixMat, price is not the most important thing, the most important thing is who takes responsibility for the part and the fact that it is delivered on time. What I do not want is to have lots of small bills of €5k or more*". Together with the entire Project Team, he agrees that it is necessary to "*go for it with MixMat because they are cheaper... for them, they are the first choice*".

He also announces that the internal project for the development of the G insert has been deprioritized in the R&D center involved, as it has no further money. This part will not be finalized – i.e. it will not be approved in a crash test – before being included in the DELTA project before the beginning of the following year – so there will be an additional risk during the crash test of the DELTA system.

The Innovation-Purchaser asks MixMat if preliminary studies and the quality of a specific element of the tool have been taken into account in the

8 For a toolmaker, modifications can constitute an additional turnover, which can represent on average 20% of the total cost of the tools [GAR 96, p. 61].

preparation of the last quotation. So, a new version of the quotation is then put together and sent by Mercury.

During a subsequent exchange, they are asked if the guarantee of the tool and the setting up of a confidential zone for carrying out the tests has been taken into account. After confirmation, these two points are integrated, without additional costs. All the points of DELCAR's specifications are then found in MixMat's quote and then approved.

The Project Manager can and has to launch a formal purchase request, i.e. in the DELCAR information system, so that it is possible to move forward with MixMat. The Innovation-Purchaser informs the Process Manager of such a necessity.

> **From: Asclepius**
>
> In order to advance in the purchase of mold for the DELTA project, you should launch the purchase demand, specifying in particular all the specifications that will be necessary for the good reception of the mold. I attach to you the data that we worked on that allowed me to advance from a Purchasing point of view and to approve the points of these last days with MixMat.

This purchase demand allows Innovation-Purchasing to initiate the negotiation of the contract itself. The Innovation-Purchasing Director sends to the Innovation-Purchaser an example of a letter of appointment[9] he used eight years earlier for investment purchases.

In response to this email, the Process Manager answers:

> **From: Mario**
>
> Please can we firstly share this document with the Project Management and launch the purchase demand together.

9 The letter of appointment corresponds to a simplified contract between DELCAR and its supplier. Compared to the traditional purchase procedure that results in the issuance of a purchase order, it accelerates the formal commitment of DELCAR while providing an additional degree of accuracy with respect to an acceptance of the quote. It also makes it possible to negotiate the terms of the partnership between the two companies.

To which the Project Manager responded:

> **From: Zeus**
>
> The main goal for me is that we receive the mold and therefore activate the entire payment only if the parts are "good definitions". Which implies: having a relevant technical criterion (thickness in such a place? Surface appearance of the main part? ...) and having a means of control ideally at MixMat.

In response to this last message, Mario amends the document that forms the final specifications and specifies the elements to be controlled.

The following week

Mercury, MixMat's Business Manager, leaves a message for Mario asking for an update.

Mario fears losing MixMat "*if we do not move forward*". He completes the specifications sent by Innovation-Purchasing: delivery dates in July and delivery locations. He worries about schedule issues:

> **From: Mario**
>
> How do we manage the delays with this project? it would be necessary that the mold maker begins to analyze the DFN to make us his requests for possible evolutions. This phase is unavoidable and can take 2 to 3 weeks, we cannot wait for approval from the monthly innovation committee[10] to start working with him, if you want to stick to the schedule. As you requested, I can send the DFN to the mold maker, but I suppose he will want a commitment from DELCAR before he starts working.

He asks Innovation-Purchasing for help on which approach to take.

The Innovation-Purchasing Director answers:

10 Referring to the passage of a milestone that allows you to move from one phase of the innovation project to the next.

From: Orion

Asclepius exceptionally wrote these specifications (which he will complete with your latest information) specifically considering the elements that the Project team (+ the supplier) could provide and to limit any future amendments. Asclepius will check that the latest elements added to the specifications are included in MixMat's latest offer. They must then sign the corresponding purchase order. After receiving this signed purchase order, Asclepius will finalize the negotiation before asking the non-production purchasing to place the final order. To save time now, Mario, you can have a purchase demand signed by adding to the amount of the last offer the estimated amount (upper range) of the potential costs of the last needs not included in the last MixMat offer.

The Project Architect then requested the Director of Innovation-Purchasing to support the project in this process:

From: Hephaestus

We now have a successful product definition. I use the term successful to mean that it corresponds in every respect to the features we wanted to implement on this product.

The next production step is 3D finite element analysis. We are very confident about the positive result of this step because we have been working for 2 months on calculations with a surface model very, very representative of this 3D volume definition and which allowed us to highlight all the strong and weak points that existed in the first definition and of course to fix it. In conclusion, we are very confident in the 3D existing to date.

Of course, this definition has been worked on with our industrial experts (spoils, demoulding... and other stuff that I do not even know). But we all know that it would be the first mold that would be launched without modifications at the request of the mold supplier. The development of the mold MUST begin on March 20 to meet the deadlines already contracted, so we must ABSOLUTELY begin the definition review with the selected mold maker.

Please use your influence to select and inform the mold maker of the imminent order at the earliest convenience so that the latter can start as early as possible. If there are any modifications (and everyone's experience proves that there will be some) they must be taken into account in the calculations and approved by 20th March. Thanks for your support.

The Innovation-Purchasing Director answers him:

From: Orion

I am very happy to hear that you are very confident about the current 3D. What do you mean by: "we must ABSOLUTELY start the definition review with the selected mold maker"? What definition are you referring to? Parts, molds both of them with an iteration process ??????? I am not an expert on the subject. To limit future amendments, we must now define what we know before placing the order by completing the specifications. If necessary, we can organize additional meetings with the "presumed" finalist, knowing that to date we have only a confidentiality contract with them.

He then goes to see the Architect with whom he agrees that it is necessary to organize a meeting with MixMat to "review" the 3D definition of the DELTA system for "possible modifications to this part given the mold and injection constraints according to the know-how of the mold maker". This 3D definition has to be sent to MixMat through a secured electronic exchange platform.

The same evening the Innovation-Purchasing Director contacts the Innovation-Purchaser to give him instructions on the relationship with MixMat: he has to send MixMat this DFN clearly specifying that it is confidential, as well as the updated specifications, asking the Business Manager to check that it corresponds to the last MixMat offer.

He also has to organize a meeting for the following week with MixMat, the Project Architect and the Process Managers. "The idea is to have MixMat work for free for us if possible. Tell him that it will be for us to comment on this part".

At the same time, the supervisor of the Process Managers indicates by email that:

From: The Process Director

It is out of the question to anticipate an order without agreement from the monthly innovation committee. If the date is too far away, it is necessary to organize a management meeting to make a decision.

To which the Innovation-Purchasing Director responds:

> **From: Orion**
>
> Purchases will not place any order without a signed purchase demand. For the rest, if the supplier can refine his offer with more elements on the subject in his possession (confidentiality agreement is signed) and already provides us with some initial feedback this should allow the project to gain some time to move forward.

Mercury of MixMat is not able to participate in a physical appointment in the days that follows. He offers either a conference call or a meeting at a trade show within 15 days.

He resends his cost breakdown and the annotated specifications: each line is approved by a handwritten "*ok*" with the exception of three lines: the date of the first test which has to be done 18 weeks up from receipt of the order, and two questions. To his email, he attaches (as for each of his emails including a quotation) MixMat's general terms and conditions.

The following week

The Project Manager asks the Innovation-Purchaser in the last email to the Innovation-Purchasing Director to confirm if it is no longer possible to place the order when the DFN will almost not move: "*it will be first and foremost a question of whether one can remove from the mold or not; and only the mold maker who can answer that*".

The Innovation-Purchasing team comes back to meet him and evaluates with him the additional cost of the specification lines that are not included in the MixMat last quotation. This makes it possible to evaluate the cost to be indicated on the purchase demand.

The same day, Mercury contacts Mario by phone. Only technical questions are discussed: Mario "*certainly does not say anything that could make the supplier believe they have been selected*". He also solves with Mercury the implementation terms of the specification lines that were not included in the quotation.

The day after, he sends him an email to confirm having sent him the latest version of the DFN and asks him to "*take note and get back to us regarding the feasibility*". Mercury answers the following week by emailing the Innovation-Purchaser, indicating that he will discuss the subject with Mario at their scheduled meeting at the trade show the week after.

The next day, the process for listing a new supplier is launched. The MixMat supplier was listed under its old name but was no longer active – it has to be recreated. To finalize the purchase demand, a meeting is held between the Project Manager, the Process Manager and the Innovation-Purchaser. The Project Manager starts the meeting by explaining that he will not launch the demand until it is approved by the Monthly Innovation Committee "*at this present time*". He mentions the risk of seeing the project stopped where it is; VROUM purchasers are studying its profitability.

The following day, at a meeting on the DELTA project with VROUM, the digital definition used for tool quotation is put on hold. Then Mario explains the tasks to be done at DELCAR to achieve the specification parts not included in the quote. No work is done around the purchase demand at this meeting.

The following week

The signing of the purchase demand is postponed until the next monthly Innovation Committee: 15 days later. At the same time, Mario asks Innovation-Purchasing how to deal with scheduling issues:

> **From: Mario**
>
> In order to move forward with MixMat and to plan their technical department activity, they wish to have a DELCAR email indicating if they are selected for the development of the mold. From there, they can plan the hours and can start working with us.

This is not possible: this type of email is a commitment by DELCAR, according to the Innovation-Purchasing Director.

The day after, the Material Expert, Dionysus, asks the Innovation-Purchaser to request a 3D definition of the mold from MixMat to check that it does not break. "*It's a risky mold. Even if the toolmaker knows how to do his job... Naturally the mold slides around, which increases the risk of the remoulded areas bursting... and who will pay for the repairs?*"

For the Innovation-Purchaser, if it appears in writing, the risk will pass on the side of DELCAR. He takes the advice of the Process Managers, and they share it. On the same day, Mario speaks to Mercury on the phone: as MixMat does not have at least one email, he cannot plan anything.

Mario does not understand why we do not tell MixMat verbally that they are selected and negotiate the price: "*it's the job of a buyer it seems to me. And in development, we did not wait for the purchase demand to start working with suppliers. Sometimes it was a little rogue. But you could tell him, it's like a story he plans. In the meantime by waiting for the monthly innovation committee, we lost 1 month*".

The Innovation-Purchaser asks what attitude to adopt if the project is cancelled. For Luigi: "*we have not made any commitments. They move forward at their own risk and they know that*". Meanwhile, the Innovation-Purchasing team continues to put together the letter of appointment in order to be able to quickly contract the supplier. This can be done as soon as they have an approved purchase demand – and therefore without waiting for the ordering process.

Therefore, the Innovation-Purchaser asks the Director of Innovation-Purchasing about the possibility of finding new suppliers due to the Project-team's progress with the DELTA[11] project. To him, this seems complicated due to scheduling issues. MixMat also have the advantage of being geographically close. He therefore asks if MixMat works in the automobile sector. Upon learning that their main clients are in the aeronautical sector, he fears that their sales portfolio is not solid and that they have no immediate needs for DELCAR's business; their strategy could be to progressively move towards the automotive industry, anticipating the time when aeronautics will be less profitable.

The same question is then posed to the Process Managers. They therefore say that we could always say there are certainly other suppliers who work well, but it is not just a mold we are looking for. Uncertainty about the main technical difficulty is not raised. They consider themselves "*on the same*

11 This request from the researcher/innovation-purchaser is one of the evaluations of the importance of MixMat for DELCAR, which was done throughout the observation.

wavelength" as MixMat in that "*we both want to work towards (this type of technology) and discover how it works*". They remember that, to approve the mold, they must be able to produce approximately 30 good quality parts. Furthermore, in order to finalize the letter of appointment, the different due dates for the supplier are rescheduled.

Mid-March 2013

At a Trade Fair where the DELTA project's Process Managers and MixMat's Business Manager must meet, the Innovation-Purchaser meets the latter who tells him that he must see them this afternoon. This is because, with the latest DFN and the breakdowns and sections they added, when discussing this with his team and comparing it with the previous DFN in terms of costing, the quotation for this DFN increased by 25%. During the afternoon meeting, they agree that the DFN used is obsolete. All points raised in the meeting are then the subject of a summary email with the correct DFN.

The Innovation-Purchaser and the Process Manager then study the specifications which could change or lead to additional costs; these appear minor and limited to changes to the position of the H-elements inserted into the main part as well as design changes that would only result in some more minor manual work or changes that can be made with another supplier in the second phase. The schedule is pushed back a few days, meaning the first supply of parts will be available at the start of DELCAR's annual summer break.

At the start of the following week, the Process Manager presses MixMat to send him an updated quotation following their last meeting – the Innovation-Purchaser has been copied in, but has not been made aware of the initiative:

> **From: Mario**
>
> In order to start the mold manufacturing with the least delays, we plan to have the demand signed by our superior on Wednesday morning during the internal technical review. For this, I will therefore need the updated quotation for Wednesday morning at the latest, as discussed last week.

The Director of Innovation-Purchasing responds to him:

> **From: Orion**
>
> Why has standard procedure not been followed? Why has this message been sent directly to MixMat ? This will not help Asclepius' role nor limit delays.

He then goes to see him on the project floor. Following this, Mario says to the Innovation-Purchaser that he prefers to work with him rather than Orion, and that from today, he will no longer have any contact with the supplier; if they call him, it will be forwarded onto the Innovation-Purchasing team.

Two hours later, the quotation is sent by MixMat, taking into account the additional studies and the modifications to be made to the tool – certain modifications are not known by the Innovation-Purchaser. The additional cost in relation to the previous quotation is 20% more. Mercury and Mario talk on the phone. The day after, Mercury sends a new quotation to respond to a request that has not been integrated. The additional cost in relation to the previous quotation is now 2% more. According to Mario, modifications not known by the Innovation-Purchaser were mentioned during the technical discussions, but were not formally requested; they must be considered as options.

3.2.4.1. *Analysis: continuing the exploration, building the collaboration*

This episode is a continuation of the first exploration part except that it is marked by a shift in the balance of power between the two firms.

Progress resumes after the Christmas holidays. Exchanges increase between the two firms. Throughout this phase, the holding of a face-to-face meeting at DELCAR's offices between MixMat's Business Manager and the DELTA Project team allows them to agree on the visions surrounding each of the firms' needs and potential solutions. Through these inter-firm exchanges within this meeting and with the help of multiple intermediary objects (scale models, digital simulation, schemas) and thanks to the responses to the questions brought by MixMat, the members of the Project team increase their knowledge of the problems linked to the mold design in order to move forward with the part's definition.

Complimenting the inter-firm exchanges, internal exchanges at DELCAR, involving everyone, help them to define the subject of the collaboration. This is materialized through a document detailing the functional, technical and commercial specifications, which will serve as a basis for negotiations with MixMat. This episode is where the design of the DELTA product reaches a new technical level, which allows the project to pass to the next innovation stage. It is also a period in which the stability of the DELTA project fluctuates a lot within DELCAR.

3.2.4.1.1. Atmosphere of the relationship

MixMat refuses any fragmentation of its involvement with the request for a pre-study. Its position is to be involved in the whole project that leads to the conception of the mold and prototypes of the DELTA system, or not at all.

DELCAR starts to fear that MixMat no longer wishes to negotiate on this project, while at the same time, MixMat is considered more and more vital for the progression of the DELTA project.

The increase in interpersonal exchanges, which are both formal and informal, leads to a resurgence of increased trust.

DELCAR's positive evaluation of MixMat's ability to respond to their issues grows until the end of that period, when MixMat is considered the only potential supplier.

If there were fluctuations with the stability of the DELTA project, DELCAR's increased trust in MixMat goes hand in hand with an increase in the dependence of the former on the latter.

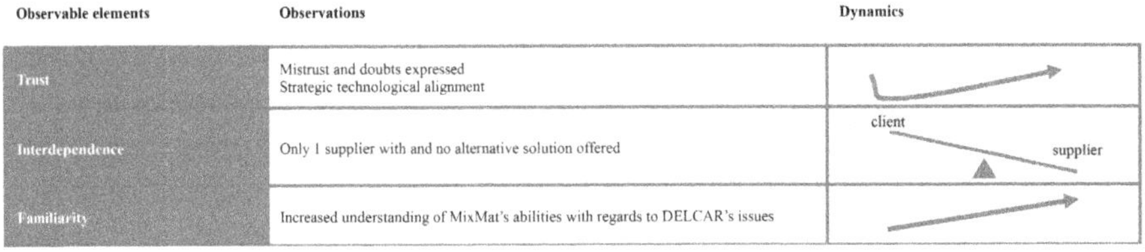

Figure 3.10. *Atmosphere of the relationship during the collaboration exploration phase*

3.2.4.1.2. Maturity of the innovation project

The relationship continues to mature during this phase: internally, via different simulations carried out; externally, via exchanges with the suppliers. The face-to-face meeting with MixMat at the beginning of this phase and the exchanges throughout this meeting have deepened the Project team's knowledge.

Following this meeting, the system's technical feasibility can pass onto a new phase; the design of the main part of the system evolved in order to take into account the manufacturing constraints which were not already known. This progression helps the team to move forward with the creation of a completed product definition.

3.2.4.1.3. Management of the relationship

The actors in the collaboration and their interactions have not changed throughout this phase compared to the previous one (Figure 3.11). Assessment work using the various experts is continued under the coordination of the Innovation-Purchaser.

Through exchanges originating from this work and through exchanges between the two firms, the future contractual elements found in the specifications, DELCAR's technical, functional and commercial specifications, as well as the elements of MixMat's costings are defined.

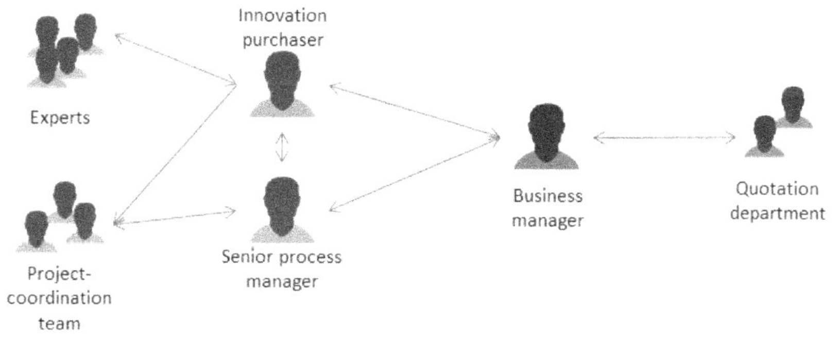

Figure 3.11. *Human resources involved in the collaboration exploration phase*

The holding of the first face-to-face meeting at DELCAR's offices with all the Project team, certain Experts, the Innovation-Purchaser of DELCAR and MixMat's Business Manager provides the opportunity to use intermediary objects and new exchange methods about these objects: scale models, future inserts for the DELTA system, 3D and 2D projections and schemas on whiteboards. It is in this context that a learning process about the relationship occurs at the most significant level.

This learning process helps specifically with the design process of the DELTA system as well as with the decision to select MixMat and to launch the internal purchase request process.

During this period, MixMat puts in place an influencing strategy which involves highlighting its need to block its resources into the schedule to be able to fulfill DELCAR's request on time; however, this blocking can only be done with a commitment to do business.

In order to respect the project's schedule and to quickly commit to MixMat, a unique process is prepared by Innovation-Purchasing: it is to submit to the supplier a letter of appointment which contains the elements of the specifications and which is negotiated with them, between the moment when the purchase request is official and that where the purchase order is effective. The continuation of this process is suspended as the project passes the following gate on DELCAR's innovation management process.

3.2.4.1.4. Performance of the relationship

During this phase, the performance of the relationship continues to be assessed through the convergence between the different requirements put together by DELCAR (different versions of the specifications and requests for the supplier to conform to DELCAR's response format) and the responses given by MixMat. This convergence increases during this phase; MixMat responds in accordance with DELCAR's requested format.

Furthermore, DELCAR's objectives regarding the DELTA project fluctuate all throughout these two and a half months. The project must be launched, it must be stopped, it will be launched, but maybe not. Its objective is not stable when creating a realistic scale model nor when approving a new technique, for the crash test and/or manufacturing process. Yet, these fluctuations are not apparent at an inter-firm level. The relationship continues to be built despite the above.

In addition to this, during this phase, the relationship is marked by MixMat's proactivity, who agrees to talk with DELCAR regarding technical problems and propose solutions which will contribute to the progression of the DELCAR innovation project. All of this confirmed MixMat's credibility in the supplier selection process. This proactivity was particularly noticeable during the physical meeting which marked the beginning of this phase.

3.2.5. *Passed gate opening the door to collaboration*

Mid-March 2013

The Monthly Innovation Committee meeting is held. The Project Manager presents the DELTA project's state of progress in order to pass the next gate, the approval of the concept. The Director of Innovation opposes it. For him, it is not possible for them to go any further if VROUM is not in agreement: "*even if they pay for the prototype, if they do not agree with the manufacturing scenario, we cannot go much further. It would not make any sense for them to pay for a prototype without thinking about industrialization. We must review this with their purchasing organization for the production and not for R&D.... we are always in pre-development*".

For this, even though VROUM wants a new design that is possible with this type of innovation, the gain in weight is not significant enough to justify a change in technology. "*My point is not to define a target price but to agree on a manufacturing scenario*". In the same meeting, the senior managers of the Program Development department request that a review be made of the solutions to the project's technical difficulties, which do not appear sufficiently solid at this stage.

The project gate is not passed entirely.

In order for the project to progress, i.e. to pass the gate definitively, an agreement must be made with VROUM followed by a technical review. This is after a dedicated review with experts in their specific roles.

The Project Manager unofficially summarizes to all his team and Innovation-Purchasing the following:

> **From: Zeus,**
>
> Gate 2 has conditional approval, provided that VROUM agrees to order the complete system and not the innovative part alone for the VROUM project 2018 ➔ there will be no launch of prototype tools without this agreement.

This suspension means a minimum delay of a month for the DELTA project if it goes ahead. The Director of Innovation-Purchasing asks the Innovation-Purchaser to inform the two suppliers that the project has been delayed due to internal reasons. The latter tells MixMat to release the planned teams for the moment – four days later, a MixMat staff member is positioned to work on the subject; they are also told that the last price increase was one of the reasons for the suspension.

Beginning of April 2013

After 15 days, MixMat's Business Manager wants to speak to the Process Manager over the phone as well as by email. And only to him. He wants to ask him for an update on the project. With no luck. Mario does not call him back.

He sends the information to Innovation-Purchasing and asks them what answer he should give with regard to the progress of the project "*I guess he wants to plan his tasks for the coming months*". With their approval, Mario responds to Mercury that the project is on *stand-by* for the moment.

A month after the Monthly Innovation Committee meeting, the future of the DELTA project is now seen in a different light by the Project Manager: VROUM would consider extending its need to other programs by following the industrial scheme in line with DELCAR's wish. It could resume in a month's time.

Nevertheless, for the Project Manager, it seems unwise to start production for the DELTA project on "*a mass production level*". He also worries that DELCAR's general management will refuse to engage in the production of such an innovative system.

At the same time, the technical review has approved the DELTA project's technical solutions and the green light has been given by the experts and their direct superiors. The Director of Innovation-Purchasing is less positive with regard to the project's next steps: sales are not good and general management "*even stops projects which have been signed*".

Following the technical approval, the Process Manager asks the Innovation-Purchaser if it is possible to ask MixMat to start the studies. The Innovation-Purchaser responds to him that MixMat would not want to carry out the studies without developing the tool and he suggests that Mario finds a solution with the Director of Innovation-Purchasing. This solution can be found through a letter of appointment "*focused on studies with a potential for mold purchasing*". This letter of appointment can be put together by the Project team following the approval of the purchase request.

At the same time, the Process Manager's direct superior suggests moving forward with this subject by sending the following message:

From: The Process Manager's direct superior

To: Director of Innovation, Directors of Development who were present in the technical review, the Project Manager and the Director of Innovation-Purchasing

...that's why I strongly recommend raising a purchase request AS SOON AS POSSIBLE only for the pre-study of the DELTA mold for the amount of 50k€. Even if in the future, we do not want to order the mold, we can stop and still have finalized the feasibility of integrating the G insert. Let me know your thoughts and decision.

15 days later

At a presentation meeting with DELCAR's general management, they announce their desire to limit innovation projects that call on the technology of the DELTA project. This should not overshadow the technology corresponding to DELCAR's industrial tool. They should stop and inform VROUM of such projects for these programs. This raises the DELTA system's target prices and reduces the chance of passing the gate with VROUM.

The project remains suspended.

As for the Project team, they had planned to launch the tooling in June – five weeks later – in accordance with VROUM's project schedule, which does not allow them to raise the order until the end of July. Due to budgetary restrictions, the order must be raised in the first semester.

3 weeks later

A VROUM vehicle has been equipped with a DELTA model and it has been well received. However, VROUM has noticed that the project has been suspended and asks the DELTA Project Manager about this. Also, Zeus asks for a definitive decision to be made regarding this project. At the same time, he asks about MixMat's interest in working with DELCAR; he fears that they would not have enough resources to dedicate to the DELTA project's tool development.

3.2.5.1. *Analysis: suspension of the relationship with MixMat*

The DELTA project has reached a new level of maturity, the project has been authorized to move to the next gate of DELCAR's innovation system management.

The passing of the other gate is suspended by the Innovation Committee: the reliability of the economic model with the automobile manufacturer and certain technical solutions are asked to be tested. For more than two months, any exchanges between DELCAR and MixMat are suspended.

DELCAR's doors are open to a collaboration with MixMat. But they do not close during the two and a half months of this phase.

3.2.5.1.1. Atmosphere of the relationship

Qualifying the atmosphere of the relationship during this suspension of exchanges phase is paramount. On the one hand, both firms want to work with the other. The understanding of one another continues without interacting: DELCAR has a clearer idea of MixMat's planning management as the latter is testing supplier management in uncertain situations.

On the other hand, there are no exchanges – trust in the next stages of the relationship is suspended; it can be assumed that the trust is decreasing as the wait increases.

3.2.5.1.2. Maturity of the innovation project

Following the presentation of the project's progress to the Innovation Committee, the project cannot pass the gate if new investigations are not carried out.

MixMat is informed of the project's suspension. Then, all contact between DELCAR and MixMat is stopped.

Commercially and technically, new guarantees are researched and established during this phase. The Innovation Project has matured solely within DELCAR.

3.2.5.1.3. Management of the relationship

During this period where DELCAR works to pass the milestone, there are no exchanges between MixMat and DELCAR after the announcement of the suspension of the project.

MixMat's Business Manager tries on one occasion to contact DELCAR's technical counterpart, with no response (Figure 3.12). Within DELCAR, technical and commercial coordination continues, specifically at a senior management level (those outside of the project in order to pass the gate).

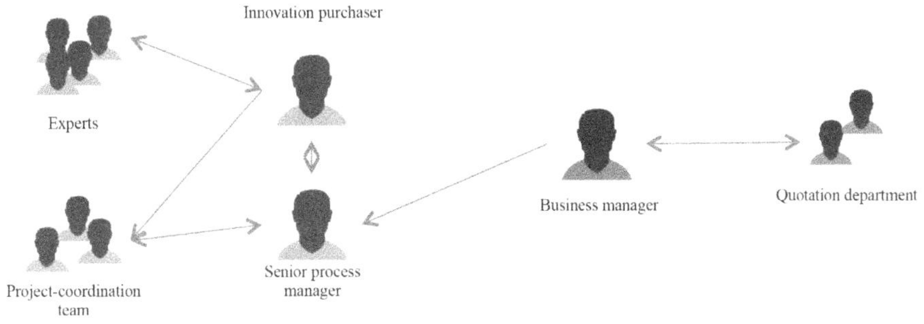

Figure 3.12. *Human resources involved during the suspension phase*

If there is no management of the inter-firm relationship, means are put in place within DELCAR to prepare for an eventual resumption of the relationship. By anticipating the passing of this gate and in order to accelerate commitment to MixMat and the start of the tool design, the process of putting together a letter of appointment is proposed by Innovation-Purchasing to the Project team, who agree with this. The letter of appointment is prepared for Innovation-Purchasing based on the latest version of the specifications approved by MixMat.

3.2.5.1.4. Performance of the relationship

During this phase of no interaction, the performance of the relationship could be considered low or non-existent: there is no convergence of the objective between the two organizations, and coordination efforts on MixMat's part are not successful.

3.2.6. *The negotiation of an agreement*

Beginning of June 2013

The DELTA project seems to have resumed: management at VROUM and DELCAR have started discussions again on the next stages of the project.

A week later, the Process Manager asks the Innovation-Purchasing department to resume the project. The Innovation-Purchaser sends the latest specifications to him. At the end of the week, the news comes out, officially announced by the Project Manager to everyone involved in the DELTA project:

> **From: Zeus**
>
> I have the pleasure of announcing to you that we will launch the tooling of the DELTA-vehicle β for VROUM. Nevertheless, I am waiting for the green light internally so that purchasing can officially raise the order before the end of the month. I do not doubt that this new step will be a success, one which will demonstrate, once again, this product's strong potential. In particular, I want to thank you all for your involvement and your motivation despite the uncertainties we have faced lately...

The purchasing process is relaunched. The specifications are updated by the Process Manager as well as the digital definition by the Project Engineers. The Project Manager tells the Innovation-Purchaser that he is "*watching with Orion how we weight the payment terms following the main milestones identified for the tooling development... and renegotiate the price if possible*".

At the start of the following week, the DELTA Project team gathers with the Director of Innovation-Purchasing. The key dates of the project are fixed: end of the tool study, receipt of the first parts and so on.

Following this meeting, Mario contacts Mercury by phone and then emails him the new version of the specifications, the project schedule and the part's digital definition – there have not been any exchanges upstream with Innovation-Purchasing on this version of the specifications. The schedule sent is the Project team's schedule and includes information related to the payment milestones that MixMat put forward in the meeting – these are linked to the project's requirements with regard to the development of different parts of the test.

Mercury from MixMat confirms in writing that the modifications do not affect the price:

From: Mercury

To: Orion

CC: Mario, Asclepius, Zeus, Apollo

Following the various discussions, I confirm the following points to be noted:

We have made an overlay of the current CAD definition; it does not call into question the initial planned costing

– Start of the studies on 08/07

– Pre-study tooling acceptance on your site in the presence of your technical team on 22/07

– Launch of the supply of wholesale steel in S30 for delivery in our factory in S35 at the latest

– Start of the rough machining in our workshop in S36

– Approval of a proper definitive study for manufacturing on 09/09

– Finalization of studies at the end of September

> – Finalization of tooling in S50 with test (…) immediately after
>
> Payment terms:
>
> – 30% for the order upon receipt of the proforma invoice
>
> – 60% + 100% VAT upon delivery of the tooling to 45J, end of the month
>
> – 10% guarantee retention upon the delivery of the produced parts, not exceeding 1 month from the delivery to 45 days, end of the month.

To this email, he attaches the quotation following the DELCAR model, the quotation following MixMat's model and MixMat's terms and conditions.

The Director of Innovation-Purchasing, the Innovation-Purchaser and MixMat's Business Manager then talk on the phone to agree on the schedule. The latter starts by confirming that the modifications are insignificant and do not affect anything. The studies can start in two weeks.

After the first phase of the pre-study, before August, a meeting will be organized between the two firms' technical teams to approve the options taken:

> **From: Mercury**
>
> We are coming with the technician from the Technical department. Each one will give their opinion. We will make a summary and we will work on the definitive study until the end of September. We will also launch the supply of metal at the end of July to have it when we return from holiday. At the beginning of September, we will start manufacturing. By the holidays the supply will be done… we will be at the end of the study and around the 10th of September we will approve the definitive study (the following week). And I will have a GO for the final part of the manufacturing. It leaves 2 weeks for the wiring and the nomenclature. I will make you a very detailed schedule.

It is agreed that the first tests and deliveries of the good-quality parts will be carried out before the end of the year.

From: Mercury

Week 50, we launch the campaign and if there is a problem with the mold we will directly intervene with a technician. We block the press for 3 days. The client, you, comes on the second day.

The Director of Innovation-Purchasing asks Mercury to send an email to confirm the new elements of the schedule:

– "*Ok, I will write to Mario*", Mercury responds to him

– "*No to Asclepius and me…*", The Director of Innovation-Purchasing clarifies. "*For prices we will see after but verify everything. What are your payment terms?*"

– "*30% for the order, 60% on the delivery date, 10% guarantee retention blocked until the approval of the parts*"

– "*Send me your stuff. We will see after what we will do. Let's be clear about the schedule and the content and we will see later what the best for both will be*"

– "*The 30% is for my supplies. 60% 40 days end of the month*".

Following this conversation, the Director of Innovation-Purchasing calls the Project Manager to feed back to him and announces that he will lock lateness charges into the contract. He says that the Process Manager should not be told anything and that he should "*only be involved on a technical level*".

Once the project management financial controller's signature is obtained at project level (nos 2, 3 and 4 top managers of DELCAR have already signed), Innovation-Purchasing will put together a letter of appointment, and the order will then be raised. For the Director of Innovation-Purchasing, Mercury believes that the business has been won by MixMat. The Project Manager hopes that the management controller will sign by the end of the month and release the 30% of the payable order as "*they don't have the budget in June*". The direct superior of the Project Manager is responsible for releasing this signature.

The Director of Innovation-Purchasing then sends the following email to MixMat's Business Manager:

From: Orion

Following our last telephone interview, we have noted your acceptance of the following conditions of our offer:

– The current CAD definition does not call into question the initial planned costing; it is part of the technical specifications and the complementary technical document that Mr Mario sent you yesterday by email

– Start of the studies on 08/07

– Presentation of the pre-study tooling to (DELCAR) in the presence of our technical team on 22/07

– Launch of the supply of wholesale steel in S30 for delivery in our factory in S35 at most

– Start of the rough machining in our workshop on 5th September

– "Suitable for manufacturing" pre-definitive study on 09/09

– End of detailed studies at the end of September

– Finalization of the tooling in S50 with a test at (external) test center

– Delivery of 50 good-quality parts on 12th December at the latest

Payment terms:

– 10% 30 days after the order and receipt of proforma invoice

– 20% 30 days after required pre-study and order approval (metal), and proforma invoice

– 60% 60 days after delivery and receipt of 50 good-quality parts and tooling and receipt of invoice

– 10% of guarantee retention paid 60 days after the delivery of further integrated parts (February at the latest) and delivery of tooling at DELCAR's site in France and receipt of invoice

Lateness charge on the delivery of 50 good-quality parts:

– 1% of the total amount of the order per delayed day (working day) from 13th December inclusive with a maximum of 5% of the total amount of the order

DELCAR's terms and conditions are applied (conditions available at www.DELCAR.com)

I verify the need for the second mold and at the same time, you have a good look at what you can do to reduce the amount of the latest quotation you have sent".

He also asks the Innovation-Purchaser to complete the letter of appointment with the lateness charges. MixMat's Commercial Director, Apollo, also becomes involved. He asks if lateness charges apply in the case of the first delivered prototypes, without specifying that the parts are good-quality, because "*this is a development project, we cannot guarantee good-quality parts in the* ***first test*** *(2 tests are planned in the response)*".

However, for the Project Manager, what is important is "*to provide 100% OK-quality parts in week 50. This point must be judged/approved by DELCAR*".

Then follows an email exchange between the Director of Innovation-Purchasing and MixMat's Commercial Director:

From: Orion

To: Apollo

Could you please review the latest changes to MixMat's position? For us, the deadline of the first good-quality parts is essential. We need a guarantee regarding this. Take into account the previous technical exchanges between MixMat and DELCAR, these did not seem to pose a problem.

From: Apollo

To: Orion

The notion of good-quality parts is relative, especially concerning an R&D project, the most linked to lateness charges. As stated in our previous technical exchanges, MixMat will make every effort with its skills and knowledge to carry out the first test according to the requested schedule.

2 hours later

From: Orion

To: Apollo

As seen with Mr Mercury who called me this morning, we request that you please review your schedule thoroughly to be able to carry out the first tests as soon as possible and also to be able to ensure the provision of good-quality parts required in week 50.

The Innovation-Purchaser cannot be present at DELCAR and cannot take part in internal meetings this week. He sends his point of view to the Director of Innovation-Purchasing by email as well as the updated version of the letter of appointment:

From: Asclepius

To: Orion

In the meeting, don't forget to see what this other tool requested by Mario (and proposed spontaneously by MixMat) is. In any case, I will not raise a purchase request for €200k if the cost is €200k+10%. Firstly, because statistically there are more than 15% of changes on the mold order but also because, in terms of the rheological study, there could be more material (that's why it appears only as an option in the letter of appointment). As for other potential suppliers to replace MixMat, certainly the other supplier could possibly be suitable, but they do not have the industrialization vision, which will be key in the second phase. So ok if we only want proto-proto and issues related to industrialization will be settled by the development of a second mold later. Otherwise, I would ask the material expert who has 1 or 2 mold makers able to achieve this, and who would be of MixMat level. If my memory serves me well, there would be one that is geographically within 2 hours' drive of DELCAR. For those we have seen, there was the first supplier who could answer technically even if they were a little expensive and a little slow, and the intermediary who acts as a study and commercial office could be a bit sour (he had been nagging us over the phone to tell him why we were continuing with others).

MixMat's Commercial Director then proposes, in addition to its changes:

From: Apollo

To: Orion

First test on 12th December at the latest / tooling development, deadline to be agreed following optimization to be made / 2nd test 50 parts, see deadline above.

To which the Director of Innovation-Purchasing responds:

From: Orion

To: Apollo

Yes, but for DELCAR the need is to have good-quality molded parts. We will put on the plan the conditions of acceptance of the part to describe it as "good-quality". It will focus on:

– criteria of "standard" aspects

– a manufacturing release time, with a maximum release time of 100 seconds

– geometric criteria ~ 10 control points to meet the geometric tolerances of the plastic industry.

This will allow you to better judge the relevance of your first tests to get the first good-quality parts following the required time. We will send you this plan on Monday.

On Monday, these elements are sent to MixMat:

From: Orion

To: Apollo

In addition to respecting the specifications already in your possession, the parts respecting this plan and its annotations will be considered as good-quality. Can you confirm your 50-week deadlines for providing the first good-quality parts?

The following week

The MixMat Business Manager answers the Innovation-Purchasing Director:

From: Hermes

To: Orion

Following our various telephone exchanges and emails, here are the different steps to take into account on this project:

– The current CAD does not call into question the initially planned costing; it is part of the technical specifications and the complementary technical document that Mr Mario sent us yesterday by email

– Start of studies on 08/07

– Presentation of the pre-study tooling to DELCAR in the presence of our technical team on 22/07

– Launch of the supply of wholesale steel in S30 for delivery in our factory in S35 at the latest

Start of the rough machining in our workshop on 5th September

– "Suitable for manufacturing" pre-definitive study on 09/09

– End of detailed studies at the end of September

– Finalization of the tooling in S50 with a test at (external) test center

– Delivery of 5 parts originating from the first test on 12th December at the latest (considered as good-quality according to our plan dossier attached; the goal being to carry out your structural tests)

(this test will not make it possible to approve a cycle; it will allow the quickest possible release of the representative parts for the next stages of the project)

– Tooling development; deadline to be agreed following optimization provided (2 to 3 weeks outside of S52)

– 2nd test of 50 parts in mass production conditions respecting the cycle with automatic loading and unloading of parts

Payment terms:

– 10% 30 days after the order and receipt of proforma invoice

– 20% 30 days after pre-study approval for steel order and receipt of invoice

– 60% 60 days after 1st test (delivery of 5 parts considered as good-quality in line with our specifications and receipt of invoice

– 10% of guarantee retention paid 60 days after the delivery of 50 good-quality parts according to the specifications (February at the latest) and delivery of the tooling to the premises (DELCAR) and receipt of invoice

Lateness charges on the delivery date of the first 50 parts considered as good-quality and corresponding to the specifications

– 1% of the total amount of the order per delayed day (working day) from 16/12 with a maximum of 5% of the total amount of the order

DELCAR's terms and conditions are applied (conditions available at www.delcar.com).

I hope this meets your expectations.

Following this message, the Director of Innovation-Purchasing asks the Project Manager to approve the technical part of the updated letter of appointment. At the same time, he negotiates the payment terms by telephone: "*60% and 10% become 50% and 20%*".

The day after

The Director of Innovation-Purchasing, having received the signed purchase request, sends DELCAR's letter of appointment to MixMat's commercial team for the DELTA project. It reiterates in writing all the elements that MixMat and DELCAR have agreed on – from a technical, functional and commercial point of view.

This letter of appointment must be accepted and signed quickly by MixMat – a short deadline conditions its approval.

The first project meeting between DELCAR and MixMat's technical team is organized at DELCAR's offices four working days later.

MixMat does not get back in touch with the Innovation-Purchasing department, despite several telephone reminders.

It is necessary for DELCAR to have the signed letter of appointment returned to them. An email is sent to MixMat.

From: Orion
To: Apollo, Hermes
NO NEWS!

MixMat's Commercial Director then responds to the Director of Innovation-Purchasing by email and asks him to make modifications to the letter of appointment. This is done the same day.

The letter of appointment is then accepted by MixMat.

Within DELCAR, the referencing and order process begins. As for the H technical difficulty, two options are considered. Only one is selected, which will limit the envisaged surcharge, and there is also a two-week extension for the manufacturing of an extra tool.

3.2.6.1. *Analysis: negotiation-contractualization phase*

The DELTA project passes the next gate – with delays. Its agenda is not modified.

DELCAR must therefore contract MixMat, the only supplier able to respond to their needs, so that they participate in the design finalization of the DELTA main part and produce the corresponding mold and the first prototypes. This phase is where the contract is negotiated and where inter-firm exchanges are carried out mainly by DELCAR's Director of Innovation-Purchasing and MixMat's Commercial Director who both represent their respective firms.

3.2.6.1.1. Atmosphere of the relationship

Now the relationship has resumed. DELCAR now needs to urgently contract MixMat in order to be able to obey the commitments to its client.

At this stage, DELCAR does not envisage being able to contract another supplier unless it leads to a result where the quality is inferior to the expectations (Figure 3.13).

In order to start this phase of the contract negotiation, this position leads to DELCAR adopting a negotiation strategy adapted to the need to respect the cycle-time and the quality, all while sticking to the approved cost. It has also adapted to MixMat's need, in the absence of a credible alternative, to fulfill the objectives of the DELTA project.

A notable variation in trust is not directly observable during this episode. It is possible to observe the relative trust of the DELCAR teams regarding MixMat's technical capability to respond to their needs in terms of alternatives that have not been identified yet. The only variation is seen when there is no feedback from MixMat when asked to sign the negotiated contractual agreement. This causes restlessness at DELCAR.

Observable elements	Observations	Dynamics
Trust	Impatience due to lack of feedback from MixMat when contractual agreement must be signed	
Interdependence	1 sole supplier and a tight schedule to respect	client supplier
Familiarity	Increase in the understanding of the constraints and commercial commitment limits of both: alignment of schedule	

Figure 3.13. *Atmosphere of the relationship during the negotiation phase*

3.2.6.1.2. Maturity of the innovation project

The project is able to pass gate M2: the concept of the DELTA system is approved. A new phase of the project begins. This must lead to the approval of the DELTA product and its manufacturing process.

At this stage, the definition of the project and the subject of the cooperation have evolved since the first exchanges between DELCAR and MixMat. The DELTA product is therefore defined, the specifications of the system are fixed and the modifications will be marginal.

The need to exchange with MixMat is low on the product side but more important for the manufacturing process side. The definition of the manufacturing tool will lead to the adaptation of the final definition of the DELTA system.

3.2.6.1.3. Management of the relationship

After passing the project gate, the purchasing process is relaunched.

Interactions resume between DELCAR and MixMat. The passage to this critical negotiation phase leads to the temporary change of counterparts on both sides. Only the commercial counterparts interact, all while continuing to coordinate the previous phases with other internal actors (Figure 3.14).

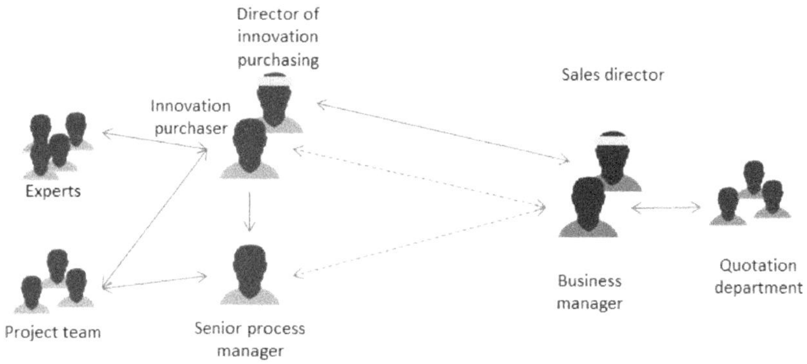

Figure 3.14. *Human resources involved during the negotiation phase*

DELCAR's Director of Innovation-Purchasing and the Commercial Director become involved primarily. They interact with each other, leading the negotiations and approving the negotiated points between themselves, relying on the letter of appointment to define a target.

The exchanges take place by email and telephone around the letter of appointment proposed by DELCAR. The negotiations focus on the commercial and administrative aspects as well as the schedule, but neither on the technical specifications nor on the functional specifications.

This phase leads to an agreement between the two parties on a definitive letter of appointment, containing DELCAR's Terms and Conditions of Sale. The listing of MixMat in the supplier management system and the placing of an order in accordance with the appointment letter are done at the same time at DELCAR following the purchasing process.

3.2.6.1.4. Performance of the relationship

During this phase, the performance of the relationship, envisaged from a relationship effectiveness standpoint, can be considered as high: the objectives of this phase are complete, and the firms are doing business together. Efficiency can also be considered high, all information and knowledge being exchanged between the two firms without difficulty.

3.2.7. *Launch of the DELTA collaboration*

July 2013

After signing the contract, the first meeting between the technical teams is held on 9th. The agenda, as defined by the DELTA Project Manager, is:

"*(1) Presentation of the tool concept by MixMat. (2) Detailed presentation and/or directly the CAD (if necessary, we can directly view files in Catia format* [12]*). (3) DELCAR/MixMat technical exchange. (4) Next stage of the schedule by DELCAR. (5) Questions/responses, put in a tracking/monitoring file by DELCAR*".

When the Innovation-Purchaser walks across the floor, he discovers that Mercury, MixMat's Business Manager, has arrived three hours before the meeting's scheduled start time.

Mercury comes to this meeting with Vulcan, the Technical Manager of the DELTA project at MixMat. He is there to "*to bring a fresh pair of eyes (...) and to launch the study*". Mercury remains the Project Manager and DELCAR's only point of contact "*from A to Z. It's how we work. You have to be able to wear all the hats*". From the beginning of the meeting, the Process Managers of the DELTA project, the Project Engineer and the Innovation-Purchaser are present. The latter starts by asking them if they need a physical scale model of the part. MixMat's Technical Manager responds that, at this stage, it is not worth it: "*We have changed phases and it's not necessary... last time we validated our concept to get close to your non-definitive definition*".

Quickly, Luigi, the Process Manager's assistant, will look for this model. It should illustrate the ongoing discussion of the external appearance of the back of the part.

12 CAD tools.

The digital definition of the DELTA system is projected on a large screen. Standing in front of this screen, the participants show and talk about the different elements which it is made up of. Vulcan highlights the points of difficulty identified by MixMat's Technical Center.

The scale model is provided. All participants stand around it and give comments. They look where the injection line would be. They talk about injection modes by looking at the different zones of the model. "*The weight is not the goal. It is not the priority*".

Mario goes out to look for a prototype from one of the two previous less complex innovation projects. He wants "*to show the finished product*". At this stage, the definition of the part is fixed; at least, there are no modifications to be carried out following this meeting's exchanges.

Mercury, with the help of a projected image, explains how the tool works to resolve technical difficulty G. Thoughts on both sides are exchanged. Another solution is presented by Mercury with the help of a fixed image. It is the "fork" solution initially envisaged by DELCAR, but it was disregarded because it require an additional operation and would not be representative of the series. Mercury is arguing for the interest of this solution.

Vulcan then mentions another "technical zone", which will pose difficulties for the manufacturing of the main part: MixMat will carry out a specific articulation in the tool, and DELCAR must refine a part of this zone. Due to reasons linked to crash test execution, it is not possible for DELCAR to make all of this part smaller. Another solution is found by selecting another point on the projection of the DELTA system.

Other modifications are envisaged to allow the manufacturing of the central part – solutions are often found by the MixMat team to limit reworkings after manufacturing – and the paperboard is therefore used to show the technical tips which can be used. A few notes are made by both parties.

The Project Manager and the Quality Manager arrive at the meeting. A discussion takes place regarding the prototype part developed in the previous project. This is compared to the projected digital definition. Other elements of the scale model are used progressively, leading to discussions on the points of difficulty.

A new difficulty regarding the integration of H insert is brought to light. A discussion takes place to find a solution. The elements of the scale model near the connection between this insert and the central piece are modified. Digital models of different elements are projected, modified and discussed. Schemas are put together on the paperboard. No consensus is reached. MixMat is asked to propose some solutions.

Mario then asks Hermes and Vulcan if MixMat will handle rheology. "*Yes*". He asks them to integrate the regulations. Then the G difficulty is mentioned again. It is necessary to establish when the final G inserts will be delivered, those which will allow them to "*approve the mold kinematic – but without molding the main part so that the process is not run-in*". These inserts are expensive in their prototype state. To reduce the cost of the elements, MixMat are asked "*How many parts do you estimate you will need for good quality?*

– "*We committed to delivering the first parts to approve the crash-test*", says Mercury, The Business Manager. "*It was scheduled. Then we will focus on the cycle-time*".

After having adjusted the technical solution to this question, the DELTA Project Manager as well as MixMat's Business Manager agree on the next review in August. The meeting must take place at MixMat premises: "*you come to us. This will allow you to come to us*". For this meeting, a three-dimensional at plan of the tool will be created and printed in order to "*approve the whole thing*". Mercury values the weight of the tool at 8 tons.

The Project Manager therefore asks them if they see "*any sticking points with regard to what was discussed this afternoon?*"

For Mercury, there is "*no specific complexity*"; and the Technical Manager "*has designed molds for 25 years; he knows what he's doing*". For the latter, the difficulty is "*you have restricted my time*".

The abandonment of an option envisaged for the H technical difficulty is thus used by the Project Manager to try to reduce the provisional schedule by two weeks. For Mercury, "*we have a cycle-time of 20 weeks. We will get to grips with it this week. That takes us to week 51. After a discussion with Purchasing, we had agreed on week 50*".

– "*Was this not in the official DFN?*" asks the Innovation-Purchaser;

– "*At first, we just integrated it into the schedule*", Mario responds;

– "*Indeed, it's the DFN which prevails*", MixMat's Business Manager confirms.

The latter, after having discussed the course of the rheological study, announces that he will discuss with the external test center at the start of September. "*We will look at the Z robot to approve the process. We will block the dates and everything else. It's us who take into account all the costs... You mention several tests?*". Two test campaigns have in fact been planned in the schedule.

Before concluding the meeting, MixMat's Business Manager poses a last question to which Mario responds positively: "*we are on a sequential hot runner?*".

The day after, the summary of this meeting is put together by Mario and sent to actors of both firms. It reiterates information DELCAR must then provide to MixMat: the main technique selected for the G point of difficulty, modifications to make to the part's digital definition, the resolution method for newly identified manufacturing difficulties and a reminder for the next meeting 14 days later at MixMat's premises. Mercury is reminded to copy the Innovation-Purchaser into any exchanges.

The meeting is then brought forward a week upon the Process Manager's request, with MixMat completely "*insisting that this shift should not call into question the overall defined schedule*"; in the confirmation email, the modification of a fragile zone is added.

At the meeting at the end of July, only Mario and Luigi go to MixMat's offices. There, the next steps of the studies are decided ("*a step-by-step process will be organized by MixMat at the test center during September: DELCAR wishes to participate in this*"), as well as the details of the rheological study. A file accompanies the summary and goes over the rest of the modifications to make, with some of them illustrated in 2D or 3D schemas.

Sends 2 hours later, MixMat's Business Manager the operating contract order for the tooling pre-study as well as its summary which states the solutions used to develop the tools and then the tests, the implications for the part's design and the elements to be provided by DELCAR. He also points out that there was an "*error on our part with regard to your last specifications which included a sequential hot runner ➔ we have planned a classic hot runner, 4 points in our offer*".

Receipt of these respective summaries is acknowledged by either party. They are not sent to Innovation-Purchasing and will be later requested from them. For DELCAR's Process Managers who are questioned by the Innovation-Purchaser 1 month later, the only salient point of this meeting is that they need them "*to get back on the right track with regard to G point of difficulty*"; the final option chosen by MixMat is not decided on at this meeting at the start of the month. They consider that "*these are things that happen, it's normal and that's why there's this type of meeting*".

3.2.7.1. *Analysis: development launch phase*

Once DELCAR and MixMat are contractually committed to each other, a first work meeting takes place at DELCAR's office with all the operational actors of each firm.

Then, during the month that this episode lasts, only DELCAR's Process Managers and MixMat's Business Manager will interact, despite Innovation-Purchasing's request to be informed of the progress of the cooperation. A second meeting at MixMat's allows the definition of the tool to progress and modifications to be made to the DELTA part, in accordance with the agreed schedule.

This episode ends with the summer holidays for staff at the two firms. By the first weeks of August, both firms are closed.

3.2.7.1.1. Atmosphere of the relationship

In this phase, in which the contractual relationship begins, uncertainties with regard to the relationship's next steps are reduced: each firm is committed to the other.

DELCAR trusts MixMat's ability to respond to its needs. MixMat trusts DELCAR's ability to pay them for their work. They believe that they are able to deliver the work in accordance with what was negotiated, this project being considered by the commercial department as a "classic" project. The power of the two firms is balanced by the signing of the contract (Figure 3.15).

Observable elements	Observations	Dynamics
Trust	Reduction of uncertainties linked to the business relationship	
Interdependence	Balance by the signed contract and mutual commitments	client supplier
Familiarity	Increase in the understanding of the resources and technical objectives of each firm.	

Figure 3.15. *Atmosphere of the relationship during the collaboration launch phase*

3.2.7.1.2. Maturity of the innovation project

If during this phase, the innovation project remains at the same level of maturity both commercially and technically, it is the subject of the cooperation that will mature through the exchanges between the two firms' technical staff.

3.2.7.1.3. Management of the relationship

This phase involves the maximum interactions and methods of interaction.

Within the first physical meeting which marks the start of this new phase, the Project team and all of the DELCAR experts are invited; from MixMat's side, there is the Technical Manager, a member of the Production team, accompanying the Business Manager. At this meeting, the intermediary objects are multiplied: diagrams, prototype, scale model, projection virtual 3D model, animation film, elements of the final product and so on. These meetings make it possible to coordinate the technical difficulties identified, as well as their understanding and their resolution methods, in order to co-create solutions.

Throughout this phase, and after this meeting, DELCAR's Innovation-Purchaser is no longer involved in the interactions between MixMat and DELCAR (Figure 3.16). He is not invited to physical meetings. He is not copied into email exchanges. He does not receive the summaries sent by both firms.

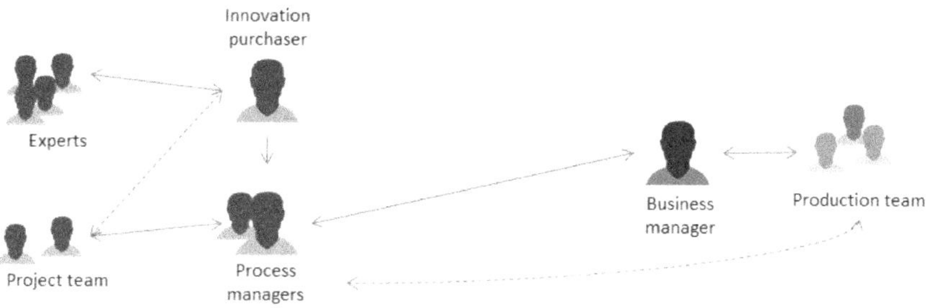

Figure 3.16. *Human resources involved during the collaboration launch phase*

It is also in this phase that, on MixMat's side, the first reviews of the contract and the precise content of the contract are seen: "*it's the DFN which prevails*".

3.2.7.1.4. Performance of the relationship

During this phase, the performance of the relationship appears at its highest. The two firms' objectives converge and are completed. The inter-firm coordination has no difficulties due to the various information exchange methods, except for the lack of interaction with Innovation-Purchasing.

3.2.8. *An urgent delay*

The penultimate week of August 2013. Return from holidays

The DELTA Project team learn today that the tool has almost doubled in weight.

Also, MixMat's Technical teams are in the process of revising their entire study, part by part, to see how "*they can get back on their feet*".

Mario waits for a call from Mercury on the subject.

When they manage to talk to each other, the latter announces that they are two weeks behind schedule. The metal used for the manufacturing of the tool has not even been ordered: MixMat's Purchasing departments have refused to launch the order before the start of the holidays, which was the only solution to be able to deliver in line with the initial schedule. Mercury explains to Mario that the delay is due to a bad assessment of the tool, each person at MixMat having "*during the holidays (...) worked on their side without speaking to each other*".

Following this conversation, Mario sends him an email to confirm that he has the correct DFN as reference "*as on another project, there was a problem related to this. And he would prefer to not have the same problem even if it was not a project he worked on*".

Responding to this email, Mercury confirms he is conforming to the DFN reviewed with MixMat's Technical Manager; he announces that he has started new metal consultations for the tooling and is working on the organization of the tests. He concludes his email by saying that, 3 days later, he has "*scheduled a summary meeting with all the MixMat staff (...) to approve the launch of the tooling manufacturing with the schedule; then the preparation of a general schedule in the afternoon to consequently be sent out*".

At the same time, a member of MixMat's finance team requests second-term payment from DELCAR, the study being delivered that same week; this request is approved by the Project Manager.

The Innovation-Purchaser requests to be sent the email exchanges and summaries put together since the meeting at the beginning of July, as Innovation-Purchasing was not copied into any written exchange since. He therefore discovers in MixMat's summary from July their announcement of a quotation error on a key element of the tool. He informs the Director of Innovation-Purchasing as well as the Process Manager who states that they had been telling them that this point was expected from the first exchanges and then in the letter of appointment. MixMat must take responsibility for this error.

At the beginning of the following week, mid-September, Mercury sends the tooling file to Mario as well as the provisional schedule detailing all the manufacturing tasks:

From: Mercury

To: Mario

CC: Apollo

As discussed last Friday; I am sending you the attached provisional workshop schedule; delivery of the tooling will be made in mid-January, with it being understood that we will make every effort to shorten this cycle-time where possible. The idea will be to be able to develop a test at the external test center in S2 after the holidays. In any event, I will keep you informed of the progress of this dossier over the 15 days with schedule updates planning and a report/photos.

Only MixMat's Commercial Director is copied in. Mario sends it immediately to the rest of the DELTA Project team and to Innovation-Purchasing.

In the hours which follow, Mario, Luigi and Mercury work by telephone on the definition of the tool: Mario and Luigi discover inconsistencies with the definition of the DELTA system (the mold must be the negative of the molded part *modulo* the constraints related to the deformations of the materials during the production of the central part). For Mercury, this is due to the fact that this definition is not the definitive one if not the one used to be able to finalize the metal order. He makes a point regarding what still needs to be done in the tooling study and proposes "*a final review starting in October as soon as the plans are finalized*". Mario and Luigi would have liked to have been informed that it was an intermediary definition straightaway, and they would not have "*wasted*" their time working on a complete verification; they are not happy.

The morning after

Mario and Luigi ask the Innovation-Purchaser if it is necessary to "*put pressure*" on MixMat.

The question then arises of the priority that the DELTA Project team makes between "*having the parts right now or having the parts in January and being sure of the quality*". A Project Engineer present on the project floor says, "*will it be of any use if the parts arrive in Mid-December if they must be stored away during the 15-day holidays? But on the other hand, it should not extend to the end of January*".

It is therefore agreed that the Process Managers should see with the Project Manager what should be done to respect VROUM's schedule. Then the Innovation-Purchaser will discuss with the Director of Innovation-Purchasing the attitude to have with the supplier and if "*we strongly pressure them*". Furthermore, it is also scheduled that the definitive definition of the tool should arrive in 10 days, when in the initial plan, it should be developed on this date: MixMat's Technical department is looking to reduce the size of the tool and cannot catch up.

The afternoon

All of the Project team and the Innovation-Purchaser make a point on the tool development project. Nevertheless, there is a deviation of more than four weeks by MixMat. To be able to adhere to VROUM's schedule, they must gain a minimum of two weeks and even then "*we are eating into VROUM's margin, who need 3 weeks to approve our data issued from crash-test before passing the gate*". The steps following the delivery of the tool output parts are reviewed.

The Project Manager will have to inform VROUM of the reasons for the delay: a complex and overly expensive tool, and a badly assessed manufacturing time by the supplier (800 hours more on top of the 3,000 initially assessed). The Innovation-Purchaser is given the job of reducing the delay to two weeks. The Process Manager is hired to carry out a weekly review of the progress with MixMat. The Innovation-Purchaser asks him to bring him copies of his exchanges with MixMat.

Upon the Director of Innovation-Purchasing's request, the Innovation-Purchaser contacts MixMat's Mercury by phone to ask for an explanation and to set noon the next day as the deadline to propose a catch-up plan:

Mercury: "*...the schedule sent is the official workshop schedule and the idea is that we will implement the mold in week 2. It's in the email. On the schedule, it is planned for week 3*".

– "*Can the schedule not be updated?*" The Innovation-Purchaser asks him.

– "*It's the schedule which was defined together with the workshop but it's an estimate. And we want to report it to week 2*", the Business Manager responds to him.

– “*There are two sides to the story*”

– “*The problem is that we have done a review with DELCAR on 31st July and then we relaunched the purchasing consultations. After the holidays, it was blocked at purchasing level because of the budget. We gathered everyone when we came back... We had taken measures to prevent delays. We wanted to informally regain it... The idea is to ensure the molding mid-S2. During the Christmas holidays we polish the tools*”.

– “*Week 4 we are crashed*”

– “*I have explained the schedule and the critical path. Now that we have all the necessary information, we know where we are going. This is the official schedule, deadline in-extremis*”.

– “*When will it be caught up?*”

– “*Day by day, roughing, finishing. I think that we can catch up. The hot runner will arrive in week 50 – it takes 12 weeks because there are shortages and it takes a lot of time. The hot runner is launched. That part is fine. We will know where we are in a week. We even approved the assembling of the inserts and the press at the test center. Everything is planned. The file is very well put together. The only mistake we had was the holiday period. And our Purchasing department cannot raise an order if we are not within the initial budget. Currently the prices are all over the place. Once you go off track, you need to review it and get back on track*”

Following this telephone conversation and using the terms of the letter of appointment, the Innovation-Purchaser sends him the following email by copying the Director of Innovation-Purchasing and members of the DELTA project:

From: Asclepius

To: Mercury

CC: Orion, DELTA team

As agreed during our telephone conversation this evening, we have noted that this schedule in pdf was an internal schedule detailing all the tasks and that we must take into account the contents of the email. The test at the test center will be carried out in S2 (instead of S50), the objective remaining at the date of delivery of 15 parts resulting from the first test. These parts shall be regarded as good-quality ... the goal being to positively conclude our structural crash-tests".

In addition, in order to be able to follow the progress of the project, we will need the updated schedule as you also proposed in your email.

We thank you in advance for this confirmation.

The next morning, Mercury sends an email to the Innovation-Purchaser and the Process Manager because he is "being chased by (his) accounting department for the settlement of the first two installments of 10% and 20% for the project launch and presentation of the pre-study tooling of 07-31". He asks them to confirm when these two invoices will be settled. The Innovation-Purchasing Director immediately asks accounting if the settlement has been made and makes sure to unblock the two payments because "everything is in order on that side (otherwise the delay would be + important!)" – an approval of the Innovation was missing in the information system. They will be paid at the end of the month.

At the end of the day, Mercury answers the email from the day before:

From: Mercury

To: Asclepius

CC: Mario, Apollo, MixMat General Director

Following our telephone discussion and your email, please find attached a summary on this project, namely:

– at the end of July as planned with your Technical department we have done a complete review of the tooling study

– launch in the wake of consultations for metals to order large blocks to start machining in mid-September

– after feedback from our suppliers we realized that we were totally off budget compared to the initial quote

– the Industrial Director blocked the purchase of the blocks and issued a meeting bringing together the entire MixMat staff (Industrial Director, Machining Workshop

Manager, Finishing Workshop Manager, Method Manager, Technical department Technician, Business Manager) ➔ the goal is to return to scope of the client quote. This meeting took place on the 03/09 after each participant had returned from holidays because the firm only stopped in week 33 and we made bearings

– at the end of this meeting, we redesigned the whole tool to reduce all the metals and to simplify the tooling as much as possible; the design was made as a series of high-production tools while the quotation was put together based on a quality material prototype tool able to develop a few hundred parts.

– I advised Mr Mario to explain the facts

– We have completely reviewed this study, and this was sent to Mr Mario on 16/09

– We redid the schedule with the workshop that I passed to Mario on 17/09

– We ordered all the big blocks which will arrive at the end of next week; everything has been done to affect the schedule of the end-of-year test as little as possible.

Summary of the financial analysis of this file made by our Technical Director:

– Changes in scope not included in our client quote, namely:

- Hot runner with shutter instead of a standard hot runner

- Added value for tunnel gates in place of simple ejector shims (visible trace punch ➔ not accepted by the Technical department)

- Added value (R metal) on the punch side in place of the S metal initially planned (possibility of graining the punch)

– Assessment of the number of working hours planned to carry out this project ➔ +900 hours compared to the initial estimate ➔ time lag to take into account this surplus of work

– Estimated balance on this file ➔ loss of 25% compared to the selling price

Conclusion:

– The schedule shift is justified in relation to the work surplus evaluated by our Technical department

– To bring back this schedule for a fine test in S2 would mean a specific organization of each phase method with the priority being the workshop on this project

> – In the order you have scheduled late penalties that start from 16-12 that would be reviewed
>
> – Discuss with you the possibility of taking care of certain points to minimize losses on this project
>
> Looking forward to your response.

For the Project Manager, who received this email, forwarded by Mario, this implies a delay of six weeks compared to the initial schedule, when four was announced to VROUM.

The problem is that it will be partial results that are delivered to them. VROUM has two gates: the results of six crash tests in February and 24 others in May. These results are requested by VROUM a month in advance. If the delay does not financially impact the ongoing project with VROUM, there is a risk of losing the opportunity to deal with VROUM on the VR2020 program: there is a risk of DELCAR missing a gate on this new program. The DELTA project risks not being off the shelf when VROUM will seek to choose a DELTA-type system for a new VR2020 vehicle. "*The context of this innovation project is that we have a client milestone and there is business behind it*". However, the capabilities will be assessed after May.

According to Zeus: "*we must ask MixMat 6 weeks into an automobile project, is this negligible? We can, as a last resort, threaten to blacklist them at DELAUTO. Even take care of his reputation with VROUM as they try to get into the automotive sector and type α materials*".

For Mario, the additional costs upon DELCAR's request are for the tunnel gate: "*unique modification requested on 31st July, specifically to have a traceless surface in a high zone*". For him, there is only the tunnels solution that is possible instead of the ejector shims – hence an extra cost which they cannot quote for.

The response to send to MixMat is prepared within the Innovation-Purchasing department based on this information. For Orion, it is necessary to inform MixMat that DELCAR does not foresee a notable modification compared to the initial request and "*we are ready to help them if everything is delivered in week 2 by waiving any lateness charges to settle their financial analysis balance sheet made by their Director*".

MixMat's Business Manager and the Innovation-Purchaser talk over the phone. Mercury starts the conversation by stating he has "*been told off*". He then says to him:

"I sent you an email to explain what happened. I was pulled back into line by my department boss; this has had an impact on sale/purchasing levels. There is additional work, and this leads to an increase in value and a delay with the cycle-time. And I wanted to explain that.

"And at the same time, in the quotation department, there have been several quotations and along the way there has been a modification with regard to the injection – and we have not taken into account the hot block. The difference is 10k€. Today that's fine, we ordered it but the plan is to see with you how we can take them into account technically and financially. And in terms of lateness charges, we had agreed with M. Orion to start on 16. And if we face charges there is a risk that this will be deprioritized...

"It is not down to the week. It does not affect anything in the system. I want to be aligned. First of all, postpone the lateness charges to week 2.

"And on the other hand, there 3 things which go outside the perimeter: the hot block that was forgotten, – that's our error. Then the ejector shims. On your side, they want the tunnel to have no traces; it lengthens the mold by 100 mm so it needs extra steel and more machining. And on the punch part they want a grainable metal. These are the 3 things my department have pulled me up on. After we are in the target price in terms of the steel. After I've been with the others to your premises to reduce the ribbing in order to decrease machining. The objective will be to reach a satisfactory compromise to avoid erosion; this will allow us to spend a little bit and limit the deficit.

"If I pay the charge, I have no intention of delivering it for the deadline – 5 days late and we reach the maximum charge, therefore it can 5 days late or 1 month late... We put the blocks on the machines and people will work in the background whilst they focus primarily on other projects in order to reduce the hourly rate. But this will extend my delay. And if we promise to deliver it in S2... it's worth coming to an arrangement".

The Innovation-Purchaser: "*I have the same worries as you*".

– "*On 31st July I spoke to you*", Mercury responds to him. "*And now I am annoyed with what happened... it was raised to my department. The objective is to get back in line with the scope. We have redefined the runner purchasing. At the method department, the hours were recalculated and now we must reduce the hours. And on my side, I was asked to go back to the client to compare it with the initial budget. That's our internal procedure. It's ISO 9001, and aeronautical certification. The goal is to redefine things immediately so as not to penalize the client*".

The Innovation-Purchaser informs him that he disagrees with technical points 1 and 3 and that for technical point 2, he will have to investigate, and "*regarding the delay, I will see what we can do*".

– "*The 10k€ for the hot runner is my fault*", Mercury acknowledges. "*And as for the tunnel we pass 1100 to 1200. And as for the metals the 2 were grainable. And as for the potential hours, I must look with the technical office to limit erosion on certain zones*".

– "*We are also trying*".

– "*From the moment we commit to each other, we will respect each other. It's a discussion between intelligent people. On a technical level, we will make a perfect tool. We are trying to limit losses. Our directions when we have a budget making a loss. Our quotation has between 3 and 5% profit margin. If not, we go and ask our Accounts department. The idea in our organization is to anticipate everything. And then we can discuss it with our client. And after discussing with them, we will also reduce the estimated time and achieve a system which is released, without neglecting the quality... that's why our ISO analysis is to redefine things as early as possible to always be clean with regard to the client*".

– "*Point 2 to review and as for the delivery delay, I will check with the Director of Innovation-Purchasing and the Project Manager. I will get back to you at the end of the week*".

– "*We tackle the machining this week. I will put together a photo report and return everything in 15 days. Everything is going well. I wanted to explain it to you in person. I did it with Mario. We know each other well: we work together. I want to reassure him about delivery and quality*".

Following this telephone conversation, the Innovation-Purchaser verifies the highlighted questions on the technical points with the Project Manager "*the visible zone is not specified in the DFN as the Project Engineers know their digital definition off by heart, and for them, it's normal. Therefore, if this was not indicated, the objective is not to crush them but to go half and half as he thinks both firms are to blame. (...) If MixMat do not want to deliver on time or ever, we say no to the additional charges; they will be blacklisted at DELAUTO (...). In order to know how much this will cost, it is not absolutely necessary to ask Dionysos (*the most involved material expert*) who will then smoke our faces like he has done regularly and as he had told us from the beginning that MixMat would charge us back for things*".

He then asks Mario by email, while Mario is performing tests on another site, for a further explanation on this point, how it was approached, as well as an estimate of the extra cost – he suggests that he contacts the Materials-Expert or the Expert-Purchaser, both specializing in molds.

Technical exchanges continue by email between MixMat and the DELTA project's two Process Managers. Innovation-Purchasing is not copied into these exchanges.

With regard to the first results from the rheological study carried out by MixMat, they have detected some problems and proposed some solutions: "*to resolve this problem, we must increase the number of injection points or at least move them, but this is not feasible given the tooling's progress. We will do another calculation by materializing the maximum permissible injection thresholds as well as by changing the thickness of the DELTA system*".

Throughout the week, as these studies progress, MixMat shows its progress through illustrative schemas, requesting further details from the DELTA Project-team who respond to him and send him notes in return and other rheological studies carried out by the supplier of the H-insert.

The following week

After a relaunch of Innovation-Purchaser, the Process Manager emails the Expert Purchaser for the additional cost incurred on point 2 by illustrating it with a digital image of the said part. With the latter travelling, he sends the same assessment request to a Costing Expert[13] who estimates it at 1–2% of the total cost. The Innovation-Purchaser then comes to see the Project Manager on the innovation floor to ask him if he is ready to accept this additional cost. He says "*Yes. We must work in close collaboration with our suppliers; they are not there for the sake of it. He will have his 200k€ if he finishes on time*". The extension of the delivery date also seems acceptable after discussing with the members of the Project team who are present at the moment because "*what will we do with the part arriving on December 16th?*"

The Project Manager says: "*I'm okay with extending the delivery until after Christmas, as we will all be on holiday... that's fine to have the parts for week 3 to get it right the first time, because afterwards, the tests will not be carried out any sooner*". He then approves with the Innovation-Purchaser that this additional charge and the extension of the delivery date is the only aspect that will be negotiated. Then he approves with a member of the Marketing and Commercial team dedicated to VROUM that the project is still expected to pass VROUM's next for the VR2020 vehicle by presenting the results of the DELTA project – this is in order to "*know if we must put a lot of pressure on MixMat or not*".

Once this telephone conversation ends, he announces to those who are present that the project is likely to pass this gate, but that it must be able to pass it and therefore have the tests in the planned schedule. Then, everyone reviews the schedule by defining the steps through which the crash test proceeds, starting from the delivery of the mold by MixMat. The result is "*we can launch the crash-test at the start of February*".

At the same time, the Process Manager has learned of the "rumors" from a colleague from another division of the DELCAR group: MixMat would go to court with this division of DELCAR, without knowing why. The Innovation-Purchaser informs the Director of Innovation-Purchasing of this. His instructions are that, if the question is raised at the table during discussions with MixMat, the DELCAR divisions are completely separate entities.

---

13 At DELCAR, the Costing Expert has a role halfway between the Financial Controller and the Purchaser early involved in innovation projects [SCH 10] – he models all the costs associated with an innovation project during the phases of the project that are close to the development phase – after the proof of concept has been made and when it is necessary to prepare the sale to a car manufacturer.

He also requests that ongoing negotiations with MixMat must result in "*no additional costs if recent changes are made, and no lateness charges will be applied if delivery is in the second week of January*".

From a technical point of view, the finalization of the tool study progresses through exchanges between the two firms' technical teams, by email and telephone. A face-to-face meeting is scheduled in two weeks, so that DELCAR can approve it, and also in order to approve the regulation of the mold – this meeting will take place at the test center.

The following week, the Innovation-Purchaser and MixMat's Business Manager speak by phone on the date agreed the previous week (the latter not being reachable before then). Mercury starts by making a point on the progress of MixMat's work: "*the tooling studies are almost finished. The rheological study is approved by DELCAR's Technical department. The hot runner has been launched. The thermal study of the mold has been launched. The purpose of the meeting between us is to approve the environmental study – it will require the result of the thermal study. One runner is at the rough version stage, and the other is being drilled. This is ongoing; there are no problems. The meeting may be in week 43 because you must first have all the elements in hand. Otherwise this is well advanced compared to what was originally planned. We are timing things for a delivery according to the schedule*".

– "*The oral or written schedule?*" The Innovation-Purchaser asks him.

– "*Now we have to be realistic. We have three months at the most to deliver*".

The Innovation-Purchaser then announces that DELCAR is ready to waive the lateness charges if the delivery is made in S2. He then asks MixMat's Business Manager to confirm this to him in an email.

– "*Yes, with regards to my bosses I need it*", Mercury responds to him. "*And on our side, we commit to deliver the mold in week 2 without fail*".

– "*Ok*".

– "*And with regards to the hot runner. That is an investment*".

– "*If we are in agreement with regard to the delivery, there will be no extensions. After January maybe, we will see then*".

– "*We will cut you some slack with the lateness charges. And the rest is our responsibility. Send me an email with regard to this. I will pass this to my management. After that, I will approve the schedule with you in S2. After that, we are done. On my side, I asked you if for the perimeter modifications, we could retouch something. Please note that we will be able to discuss this at the start of January. After the quotation service made an error with the hot runner. It's always a problem with these quotations that drag on; this has dragged on for several months. There have been several elements and things to go through (...) regarding the department if we made a bad deal financially; we went through all the methods of how to minimize losses. Send me an email regarding this. I will send you any comments. If they tell me it's okay, I redo the schedule and we recommence*".

– "*I will send you an email on this at the end of the week*".

– "*Very good, like this we can settle*".

Following this conversation, the Innovation-Purchaser sends him an email confirmation, reviewed with the Director of Innovation-Purchasing:

> **From: Asclepius**
>
> To: Orion
>
> Hello Sir,
>
> Following our telephone conversation yesterday, we have taken good note of your new delivery schedule for the first 15 good-quality parts at the end of week S2 2014. This will be a month later than the initial schedule.
>
> With regard to the other points, there has not been any change in perimeter in terms of our request accepted by MixMat.
>
> However please take into account your email from 20th September below: DELCAR will not apply contractual lateness charges as long as MixMat stick to the delivery schedule above.

The same week, the Project Manager informs the Innovation-Purchaser of discussions taking place with VROUM, which are still ongoing and problematic due to the divergences between the two firms. At this stage, VROUM plans an annual production of 350,000 DELTA systems, while DELCAR would prefer a fifth of this due to the risk associated with this innovation.

The DELTA project could be stopped in one month.

Also, the Project Manager would like to know what could be done with MixMat if this happens and the resulting costs. Furthermore, his instruction to the Project team is to carry on working.

3.2.8.1. *Analysis: the delay and its management, a visible disruption or an epiphany of the relationship?*

The delay crisis is rooted in a period without observation and without inter-firm interactions: the month of August, when the two firms are closed for more than two weeks and are aware of employee absences during other weeks.

During this period, a lack of coordination within MixMat, and an internal reevaluation of the project, has seen a leap in the cost. As a consequence, the project is severely behind schedule. DELCAR will not be able to deliver on time.

This "crisis" episode stands out from the previous and following episode by the change in the procedure to be followed by the organizations and the actors involved.

The actors of the Innovation-Purchasing department are involved again in interactions by other actors of MixMat and DELCAR – they are once more copied into emails, reports are sent to them and they are consulted. The Directors in conjunction with the commercial relationship of the two firms are newly involved, even though they are not directly involved in inter-firm interactions (Figure 3.17).

The resolution of the crisis takes place through a negotiation carried out between DELCAR's Innovation-Purchaser and MixMat's Business Manager. They are involved in the coordination with all the actors involved respectively in each firm.

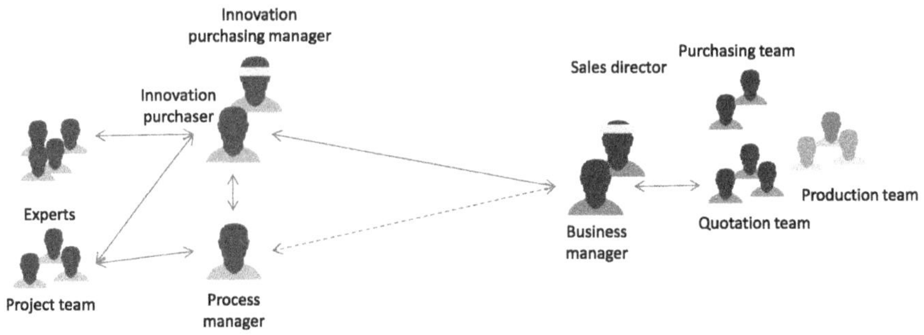

Figure 3.17. *Human resources involved in the "delay" crisis*

DELCAR's influence strategy consists of making reference to the contract, all of them being ready to accept concessions on the lateness charges if they do not surpass the planned point in the project schedule, in line with what was requested of the supplier. The lateness charges had been specified in the contract by DELCAR, precisely to prepare for this type of situation: as a negotiation leverage rather than a way to lower the final cost.

As for MixMat, its influence strategy rests firstly on the presentation of all the actions and facilities put in place in the firm in order to respond to DELCAR's needs and to resolve the crisis (mention of the ISO procedure used in aeronautics, breakdown of the events which led to the delay and those which were put in place to catch up). It also consisted of explaining the economic issues of the project internally for MixMat: a project which will result in a loss. Finally, the threat of a further delay has been used to ensure that the contractual clauses resulting in lateness charges were not applied.

The crisis has been resolved with a written agreement that the deadline will be extended and, in respect of this new schedule, that lateness charges will not be applied. It led to the implementation of new control mechanisms for MixMat's work progress. MixMat proposed the idea of regularly sending proof of progress, photos and an updated schedule. DELCAR decided to travel more frequently to MixMat's premises.

3.2.9. *A line, not so straight, until delivery*

End of September 2013

Mercury sends Mario a file for the mold's digital definition. This gives the impression that nothing has changed. He directly sends a frustrated email to Mercury. The latter calls Mario immediately and they realize that the fault is Mario's who apologizes immediately.

Start of October 2013

Eleven days after the regulation of questions linked to MixMat's delay, a meeting is held to approve the different studies between MixMat's Technical department and DELCAR, at the test center – delayed by 10 days compared to what was planned the month before, because of difficulties in finding a time slot at the test center.

Mercury sends the first tooling progress report, and a series of photos showing the mold being manufactured (Figure 3.18). MixMat's schedule detailing all the technical steps is also sent, with an end date that is missing a week, to be consistent with what has been decided.

Figure 3.18. *Proof of progress example 1. For a color version of this figure, see www.iste.co.uk/servajean/dynamics.zip*

At the coffee machine, a Project-Engineer of the DELTA project says that this project "*is not too difficult. (...) Innovation projects where there is a client involved are more complicated than others where finally we have funding and we progress. While there we do not really know where we are going, at VROUM they do not know either, therefore they are procrastinating, which makes things difficult. And when there are projects with an order behind them, it's also easier because it's on track*".

For the Process Managers, MixMat's Business Manager is "*crafty but not in a positive way. He plays around a little too much. And he tries to scrimp and save. It may be that he no longer has knowledge of the automotive industry. And yet we have not bothered him with the price or with very restrictive specifications*".

For them, "*the relationship with MixMat is all the more complicated because we have only one contact person: Mercury. When he is not there, we do not have counterparts and it's embarrassing even if we know that the other guy, the one who deals with the technical part, is a free electron difficult to manage. Furthermore, we received a mold definition that we must verify as a whole because we have the impression that nothing has changed in regard to the last time where we had requested that modifications were made. (...) Therefore, we must verify it; then call them again if necessary*".

The following day

The verification is carried out. MixMat's modifications to solve identified problems have been made.

The following week

VROUM must agree for the DELTA project to continue. "*The atmosphere is rather disengaged at the moment*" and "*it's a mess because, industrially, there are risks and we are starting to move. In fact, there is an average of 3% rejection on parts using equivalent technology in DELCAR's other division. And it's about hidden parts, but as for appearance parts, if it is brilliant or shell-like, it can increase to 20%, and that poses problems*".

Furthermore, a VROUM competitor should shortly launch a system with material and properties very similar to those of the DELTA project, so DELCAR may not be the "*first shooter*" on this new category "*but we don't have a specimen, so we don't know what has been made*".

End of October 2013

The thermic study on the tooling is sent by MixMat by email and proposes that they discuss it during their next face-to-face technical meeting. After reading this report, the Process Manager replies by email:

> **From: Mario**
>
> To: Mercury
>
> We can see that the closing force is beyond the force of press that is planned at the test center, although I am optimistic with this type of material. Can you nevertheless inquire about the availability at the test center of a higher tonnage? If they do not have one, do you have another solution if you need a superior press? We'll talk again on Thursday.

To which he responds:

> **From: Mercury**
>
> To: Mario
>
> Following on from your email concerning the closing force of the tool compared to the planned press at the test center of 15 tons and compared to the rheological + thermal study sent last week; we have planned:
>
> – A switching closing force of w tons
>
> – A pressure retention of 500 bars ➔ this involves a closing force of over 15 tons for a few seconds; however, we think that with this material we would not need this pressure retention level and this press would suffice ➔ to approve within the tests depending on the injection settings of that part.

The technical meeting between MixMat and DELCAR went well. The two Process Managers saw MixMat's installations and the test center. At the test center, they saw how the installation will allow them to go beyond their expectations with the ability to automate the installation of inserts and the ejection of the system. The exchanges with MixMat went well between the three of them: "*there are still some details to fix on the definition of the mold – the attitude was good on MixMat's part – concerning the delivery schedule, they were up to specifying the delivery date. On the 14th the mold is installed at the test center and the first adjustments are made. On the 15th the tests will be carried out*" however "*it's off the beaten track even after review*".

The Innovation-Purchaser asks Mario about the possibility of playing with the lateness charges. Mario responds to him: "*We will see. It might not be worth playing dirty with them*".

"*In any case you would not work with them again?*" The Innovation-Purchaser asks him.

"*In fact, that would depend specifically on the quality of the parts and the first parts. I usually see that the first parts do not go well so if it goes well and the quality of the mold is really there compared to other molds, that would make me want to work with them again. The difficulty with them is the lack of interaction. We cannot talk with them directly. With the other team, we send them a photo, they respond to us on the day, we send them a modification and we talk like that directly. It's difficult to work with them, in terms of the personality of the team. And they have a more global approach, they respond in bulk within 2/3 days, whereas, in the automobile industry, it is more common to make changes directly. I'm sure it is more effective in the long term, but we do not have a routine and contact is not easy*".

10 days after this meeting

A new progress report with photos is sent by MixMat (Figure 3.19), with the schedule. The adjustments are discussed by email and by telephone.

Figure 3.19. *Proof of progress example 2. For a color version of this figure, see www.iste.co.uk/servajean/dynamics.zip*

10 days later

Another technical difficulty, "*a very slight interference*", arose for MixMat after having "*monitored closely the impact*" on the tool's zone, which is represented by two schemas illustrating this zone.

From: Mercury

To: Mario, Luigi

… the modification of the draft seems difficult to achieve in the tooling because these areas are completed both in punch and matrix (welding and punching recovery matrix recovery, recovery of the opening of the gate d and the gate e); on the series parts should simply provide a coupling of the slightly inclined clip. For prototype parts; could we leave things as they are and try to mount the clip; in the case where it is strained too much, we could give it a little filing to reduce it a little bit in length.

Two hours later, a new email is sent by Mercury to the DELTA project's Process Managers to inform them of another difficulty:

From: Mercury

To: Mario, Luigi

As I told you last Monday; the dimension differences are now very important; we cannot think of assembling mechanical closure parts with such wide tolerances. I think it is necessary to wait for the return of the first hinges to approve the assembly of the key with minimum looseness. … I enclose a file showing the shim hinge assembly with the recesses planned during our meeting.

The same day, for DELCAR's Process Manager "*they must finish and deliver them to us. It's starting to get tiring*" and "*they do not listen and do not do what they've been told to do. It is the case for technical difficulty G, they are told to work on this area for 2 months and they do not do it and they want to do as they want by tweaking where it is not necessary. And now they say that the mold is done, that we cannot go back*".

Start of December 2013

Six weeks before the schedule delivery date of the first parts, DELCAR and the other division of the DELAUTO group plan to form a joint venture with the goal to produce an innovative system as well as other DELCAR parts calling on production technology of which the other division possesses the know-how and industrial tool. This solution allows "*a breakdown of a political cap on the distribution of profits*" and allows to "*unwind the situation for the DELTA project*".

A few days later, when MixMat requests a required element for the development of the parts of the DELTA project, MixMat's budget for the job is exceeded.

Also, Mercury warns Mario by phone of this overspend, and exchanges take place between the budgetary lines in order to re-establish a financial balance. He then confirms his approach to him by email:

From: Mercury

To: Mario

After the feedback of the Z robot encryption by the firm Mano and after several discussions and commercial negotiations ➔ we started from € 14025 to arrive at a discounted price of € 10000 for the development of the complete Z robot, ensuring the insert deposit functions but without the gripping part by suction pad. As I said on Friday evening; we are outside the initially planned budget of € 8000. I proposed taking the missing € 2000 in the parts testing budget ➔ initially estimated at € 5000 for the testing of 60 parts reduced to € 3000 for the testing of 30 parts. I have attached the quotation put together by Mano for this service; the Z robot will be delivered at the same time as the tooling for the tests planned on 15/01.

This email is followed by an illustrated proof of progress for the tooling.

The same week, a work meeting is held between the experts of the other division of the DELCAR group and the DELTA project's Process Managers. For this meeting, the tooling digital definition has been requested of MixMat.

The experts of the other division estimate that the test center's press tonnage was not sufficient. They "*called to ask the test center if a press of w tons was free, because the Process Managers did not want to call them*". This press would be free one week after the planned test date.

Furthermore, DELCAR's senior management wishes to stop the DELTA project at this stage. But as the client, VROUM, wishes to continue, arbitration was required at the level of the group's senior management, so that the project did not stop. Following this, DELCAR's senior management requested that the Project-team remove the difficulty linked to H point, the function of which is now technically insecure (unlike the competitive advantage that it may possibly provide to DELCAR). This modification would call into question the design of the product and may therefore stop the project.

The following week

On reading the email on the Z-robot, the Director of Innovation-Purchasing inquires whether it was included in the specifications of the initial order. This is not the case and when the Innovation-Purchaser asked the members of the Project-team how important it was, they responded that "*it is not important because it was necessary* a priori *to obtain a good part in the target cycle time*".

Yet, for the Innovation-Purchaser "*it is not impossible that they have moved towards a technical solution for technical difficulty G which mean the approval of the cycle-time cannot be done anymore... but I have not managed to have from confirmation Mario*".

In fact, the Material Expert, Dionysos, highlights that the G inserts will be put on cassettes "*which will not make it possible because it will not be at all consistent with what could be produced. It will therefore be necessary to ask after paying for a pre-industrialization study*".

However, if the other division of the DELCAR group carries out the industrialization, there is little chance that it is MixMat which carries out this study, these being blacklisted with them because "*they messed up a mold for a program*". Also, the series mold should not be done with MixMat and "*we will not benefit from the learning effect, if there ever was one*".

He also confirms that the press tonnage of the test center was underestimated. In addition, it indicates that the mold ordered to MixMat is too expensive for what is produced, and it would be necessary that Innovation-Purchasing requests the detail of the work carried out and then verifies it.

The Innovation-Purchaser asks Mario for the latest email exchanges with MixMat. Emails dating back more than a month are sent, those containing the explanations on the technical difficulties, the options taken and also those on the questions of budget.

The same week, which is a month before the scheduled date first tests, Mario, Luigi and the experts from the other division meet at MixMat's premises. They bring photos and videos of the mold in the manufacturing process. Following this meeting, Mario sends a summary to the Innovation-Purchasing department:

From: Mario

To: The DELTA Project-team: Asclepius

The other division is now in the loop/fully informed concerning the DELTA system for VROUM, they have committed to be an internal supplier for the DELTA system. Following this, a visit took place at MixMat's premises to run through/see the tool and to evaluate the RETEX product that we can introduce. One of the points that emerges with regard to the first planned moldings is that the press identified at the test center would be insufficient.

MixMat are therefore requested to look for a higher press (20 tons) in order to guarantee the release of the parts from the first molding. The deadline requested of MixMat for the first molding will not be stuck to and will be postponed *a priori* to 20/01 (instead of 15/01). There should be an economic impact (minor) and MixMat must get back to me today.

To this email, he attaches the summary put together by an Expert from the other Division, a technical report identifying the risks and actions to carry out recommendations. This report indicates that:

"*The mold is developed correctly. I confirm that the tooling is intended to approve the product, but not the general DELTA process*".

Following this meeting, Mercury puts together a quotation at the test center which he then sends to Mario, informing him that:

From: Mercury

To: Mario

…we have a quotation of 7500 € for 2 days of testing; in our quotation we had planned a budget of 6000 € for 2 tests. I propose to you that we take the budget of 6000 € to pay for this test and that you place an order of 1500 € for the add-on, as mentioned in our conversation; on our side, we take responsibility for the tooling adaptations for the assembly on zone T.

This email is sent by Mario to the Project Manager and to Innovation-Purchasing to ask them to approve this additional cost as well as a week's extension for the tests. The Project Manager raises a purchase request for the difference, in agreement with the Director of Innovation-Purchasing – for this, MixMat must send a quotation with the exact requested amount on it. Promptly, Mario sends them an email confirming the order by waiting for the arrival of an "*official*" order.

Mario then thinks that this situation emerged from the fact that DELCAR had indicated a first tonnage (12 tons) and then MixMat proposed a slightly higher tonnage (15 tons), but, in the end, it was not sufficient – in his opinion "*we would have been better leaving the subject completely open*" in the specifications rather than indicating a minimum press size.

The Director of Innovation-Purchasing asks a Purchasing Manager specializing in this type of order to approve the initial test center quotation – he informs him that the production price is three times cheaper. Later in the day, the Project Manager emails the Director of Innovation-Purchasing, asking:

From: Zeus

To: Orion

It's 1360 € more expensive (quotation from the test center's support). Now, I don't know if they are scamming us on the hourly rate or not... the question is simple: are we looking to negotiate over the lateness charges?

Then he calls the Innovation-Purchaser to ask him if they should use their delay to say that they would not pay, or not. For him, he has the money, so he signs the purchase request and sends it to Innovation-Purchasing.

The Innovation-Purchaser talks with the Director of Innovation-Purchasing on the subject, giving him the run-down on the latest exchanges, specifically those by email: the Process Manager has approved the order by email "*so we cannot do anything*". He reminds him in this instance that the order was approved on the provision of "*good-quality parts*" and not for the press rental days. Also, the Director of Innovation-Purchasing writes to the Project Manager and to the Process Manager:

From: Orion

To: Zeus, Mario

After (or possibly before) signing the purchase request, Purchasing will find it difficult to negotiate; MixMat already have an agreement with DELCAR. Can you make sure that this covers all of your future MixMat needs with regard to this mold.

If my memory serves me well, you need a second series of good-quality parts (crash-test and appearance) for February and approval of the cycle-time. Are you sure you can do everything in two nine-hour days of tests?

This email is not understood by the Project Manager – the day after, the Innovation-Purchaser explains the principle of the quotation agreement to him.

That evening

The Director of Innovation-Purchasing calls Mercury. He "*reminds him of Mario's acceptance email*"; he tells him that "*this will not happen again*".

The Innovation-Purchaser enquires with Luigi and the Project Manager on the comment from the expert from the other division of the DELAUTO group on his confirmation that "*the tooling is planned in order to approve the product, but not the general process (...)*". For them, it is for him to "*cover*" and in addition "*they have laid it on thick*" in their last email, which is then sent to the Innovation-Purchasing departement. The content of this message emphasizes that:

From: Expert from another division

To: The DELTA Project-team

…1/ Kinematic[14] prototype tooling, not representative of the series. The prototype mold approves a Product and not a Product/Process couple (insert deposit/main part ejection). 2/No robust concept mold series on kinematics. 3/ Request from us to carry out an expertise phase by expert toolmaker panel from our division, to approve a complete kinematics tooling series (product/process couple)

Then everyone finishes for the Christmas holidays – the two firms are closed during the Christmas truce.

January 2014

When everyone returns, the Project Manager returns to the last email from MixMat and makes enquiries within Innovation-Purchasing:

From: Zeus

To: Orion, Asclepios, Mario

… A detail that bothers me: Mercury speaks of 2 days of consecutive tests while we always mentioned 2 1-day sessions with 2 to 3 weeks max between the two, the time to approve the first (systems) in the crash-tests. I believe that he wants to "get rid" of our project as quickly as possible to get paid quickly. Did you speak to him before the holidays ?

Also, the Director of Innovation-Purchasing responds to him that:

From: Orion

To: Orion, Asclepios, Mario

It is the letter of appointment signed by both parties and included in the order which is binding (document attached) + new deadlines accepted last November. For the rest there is an additional clause for the "press with a tonnage of 20 tons", which is to be defined. The time and the period(s) of use are to be fixed and to be requested of MixMat according to your needs.

14 These are the movements of the different parts of the tool during its use for the manufacture of the DELTA system.

The letter of appointment is attached with this latest email.

Mario sends his wishes to Mercury, and ends it with the following message:

From: Mario

To: Mercury

...I would like to highlight 3 points in this email.

1) Today, we send out to the attention of yourself at MixMat, 20 pairs of the G insert, suitable for developing the first parts on 21/01 at the test center as well as the dummies to be used for creating the mold start-up parts (connection disk in place of the real G insert).

2) I would like to return to our discussion from before the holidays concerning the distribution of the 2 days for the molding. Our request is:

– to make the first 10 pieces on the first day as no major molding issue is known;

– to keep the second day of molding for the remaining parts once the first ones been dimensionally tested and our first lab tests have been carried out, allowing us to plan what is scheduled in week 7. We cannot make the remaining parts without knowing if they comply and if we do not need to modify the product, I think it was planned like that originally.

In addition, it would be interesting to grain the matrix before the second molding; can you give us an estimate for a Zeta grain?

3) We need to visit your premises with staff from our Commercial department + the experts from the other division of DELCAR who deal with VROUM, to visualize the mold. For this we would need to see the mold open if possible before it is delivered to the test center. We propose the afternoon of Wednesday 15/01. Please could you confirm your availability for approximately 2 hours. Also, would it be possible to have a meeting room at your premises for about 1 hour to conclude our exchanges with our colleagues from the other division?

The day after

Mario calls Mercury. For this, the email he sent is important because they will at first only have inserts G that are sufficient for the test for the crash. Before this call, the Innovation-Purchaser reminds the two Process Managers that the order indicates that it is to deliver DELTA prototype parts and not test days.

The Innovation-Purchaser is invited to this working day at MixMat, with the person who deals with the encryption of tools with VROUM. The goal is for the latter to realize what the mold is more precisely. The Innovation-Purchaser will not be able to go there.

A conference call between Mario, Luigi and Mercury is organized before this day. The question of the second wave of production of good "*crash but also aspect*" parts is discussed because it is a point of concern for the Project Manager; that would not be a problem for MixMat and "*anyway with Mercury, it's like working with an eel*". At the same time, MixMat asks to be able to present the mold realized for DELCAR during the next big trade show – this request is refused by the Project Manager:

From: Zeus

To: Mercury

Given the confidential nature of our project, it seems premature to present the tool at the show. We do not want to communicate immediately on this project.

3.2.9.1. *Analysis: next steps in the manufacturing of the tool*

Once the period of the delay is settled and negotiated, this stage corresponds to the period of realization of the tool in the premises of MixMat. While the crisis was being managed commercially, technical interactions have continued between the two firms since the return of the summer holidays, without stopping. Once settled, MixMat's job progress controls are implemented.

3.2.9.1.1. Atmosphere of the relationship

If a suspicious attitude towards MixMat has developed, the DELCAR teams remain confident in the technical capacity of the firm (Figure 3.20). The agreement found on the schedule shift has reduced the room for maneuver for DELCAR that is engaged towards the automaker recipient of the DELTA project results. During this time, when the question arose to put pressure on MixMat to force it to do something, DELCAR evaluated that it was not able to do it.

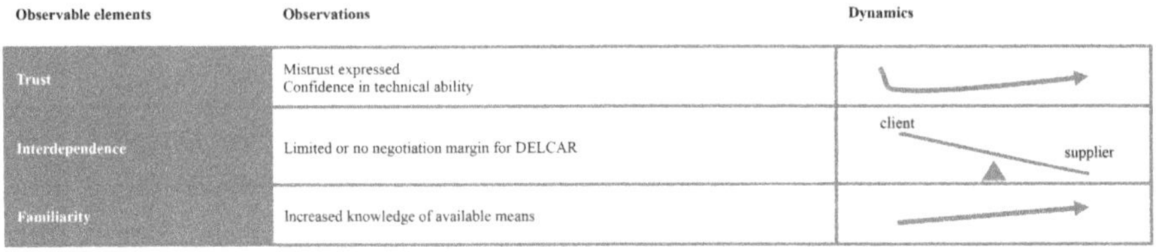

Figure 3.20. *Atmosphere of the relationship during the final development episode*

3.2.9.1.2. Maturity of the innovation project

During this episode, the maturation of the innovation project is realized through the finalization of the design of the tool, and the launch of its manufacture by MixMat. Exchanges with DELCAR contribute to furthering his knowledge of the production constraints of DELTA systems – such as solutions for point G or press tonnage.

The uncertainties of the innovation project are at this stage those which hover over its commercial success as well as the economic model which differs from the main business model of DELCAR. The turbulence caused by these uncertainties within DELCAR does not affect the relationship with MixMat during these four months.

3.2.9.1.3. Management of the relationship

Following the resolution of the crisis, the configuration of the actors involved in the projects returns to the one that was in progress during the first part of the joint project phase (Figure 3.21). The DELCAR Innovation-Purchasing team is no longer a recipient of the information exchanged concerning the relationship with MixMat. They are only involved with the project team when there are risks of additional costs (modification request or questions from the Project Manager).

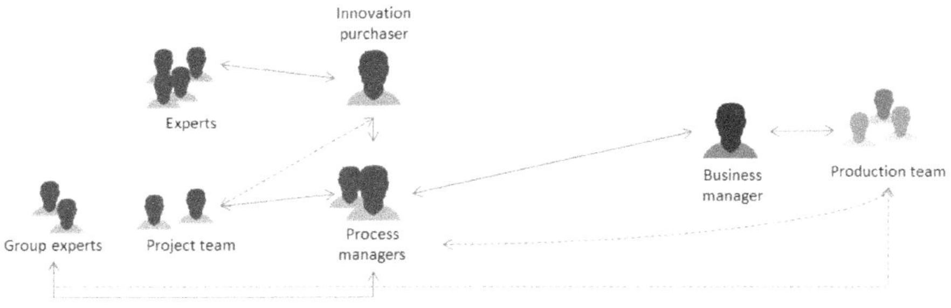

Figure 3.21. *Human Resources involved during the final development episode*

If the internal experts are solicited only by Innovation-Purchasing, "external" experts are involved during this phase. These represent another division of the DELAUTO group that could be responsible for the industrial production of the innovative product; they are called upon as experts to evaluate and prepare the transition from innovation to industrialization. They confirm the quality of the work done by MixMat.

After the amendment to the contract agreed to close this crisis "delay", another contractual device is added: an estimate is sent by MixMat following a request for modification by DELCAR. It is accepted by mail and followed by an order.

The control mechanisms put in place following the crisis are proposed by the supplier and applied: reports are sent by MixMat to DELCAR's technical contacts. They also visit their premises on a regular basis with the DELAUTO group's technical experts to check the progress of MixMat's work and prepare the prototyping stage. The modes of information exchange (schemas, digital files, physical visits, telephone exchanges and emails) as well as influence strategies (reference to contracts) remain the same as during the first episode of realization.

3.2.9.1.4. Performance of the relationship

The goals of both companies continue to converge and be fulfilled. If inter-firm co-ordination appears to be difficult from the point of view of

DELCAR, it has no difficulties: the modes of information exchange have continued to grow compared to the first episode of realization, with the multiplication of the means of control which make it possible to have a shared vision of the progress of the object of cooperation.

The difficulty of coordination comes, for DELCAR, from the presence of a single interlocutor at MixMat who answers his questions by batch and not live, which is recognized as not being usual for the technical actors of the projects of innovation of DELCAR.

3.2.10. *Attainment of the collaboration*

3rd week of January 2014

The test days are here.

The Project Manager and the Process Manager do not want too many people to attend. Zeus himself remains at the DELCAR R&D center.

At the test center, in the area reserved by MixMat for the DELTA project, the Process Managers and the Quality Manager represent DELCAR.

The first day at 3 pm

"*things are not progressing at the test center. There are problems with the slides... they have not created a part yet*".

The day after, at 10 am

A message is sent by the Quality Manager to the team members who stayed at the R&D center: "*First part completed and tested*". The Project Manager relays this later to the Innovation-Purchaser: "*A beautiful 5kg baby was born this evening 6pm ;-) The mother is doing well*". A picture accompanies these mails.

The team members at the test center then return to the DELCAR R&D center with the right number of parts.

End of January 2014 – 1 week later

For the Project Manager "*it smells bad and it's hard at the moment, the projects are all at the limit*". The CEO of DELCAR told the group's executive committee that the product family to which the DELTA project belongs is not a priority. He also said he does not want to develop the DELTA project with VROUM. One of the consequences of this statement is that in response to the request for quotation referred to by the DELTA project, namely the VR2020 vehicle, DELCAR will present a competing product that corresponds to the industrial tool of DELCAR and "*is of course cheaper*".

As far as the tool is concerned, there are revisions to be made – some are due to MixMat, and others are due to new requests from DELCAR. Mario says he is "*aware that MixMat will seek to increase the price level... there may be a problem because there were 2 days of testing on MixMat's side even if specifically, for DELCAR there were only one day of manufacturing, and again without the Z robot because it was not working... there is a risk that there will be negotiations around this*".

For the Expert Materials solicited by the Innovation-Purchaser to know his opinion on the quality of the work done by MixMat, besides the aspect that is not finalized, there are other problems and, in particular, the fact that the entry of the material to the space is not tight enough to fit properly, resulting in particular in the appearance of marks on each side.

For him, for lack of inserts G, there have not been enough tests especially with higher temperatures. At this moment, it is therefore not known how it will behave. In addition, the project is moving towards a heavier press, which will change things in the evaluation of DELTA manufacturing costs. Furthermore, the α material is not optimized; it would be possible to reduce its density. So, for him, all this also means that the crash tests will not be valid, and whatever the results, it will start again with parts that have been properly molded.

During the 15 days following the release of MixMat's mold parts and their delivery ("*a good number of good-quality parts were delivered as hoped*"), DELCAR's technical teams have specialized in "*dressing*" them in order to integrate them into prototype vehicles: dressing, adding accessories, integration with related systems and so on.

Then crash tests are launched.

The first "blank" test is not valid. There was a fault on a related system that exploded; the dissipation of energy was made at this level. The second test is the good one, as well as the following ones. The results of the tests are not communicated to the Materials Experts; their opinion was not asked on what to improve when there are changes to the tool and the design of the part, especially in order to not increase the tonnage of the press. Their opinion, gathered by the Innovation-Purchasing department, is that if the results of the first crash tests are positive while the molding is not perfect, it is because the material and the thickness of the central part have not been optimized – so there could "*finally be a weight gain that can be considered as not insignificant*".

In the middle of the trials, the monthly innovation committee is held. A collaborative project of the DELTA project family was given a go-ahead because its stop cost was double the cost of the prosecution. Also, the Project Manager is very pessimistic about this product family at DELCAR in the short and medium terms: "*there has been so much money invested in our industrial tool that there is little or no chance that there will ever be a desire to industrialize DELTA-type systems even though the first results in a 30% weight gain... even when the client asks for it*".

Whatever the case may be for VROUM's mid-March quotation request, two DELCAR systems will be submitted: the DELTA system and a "classic" system. Today, the costing service is working on the subject "*so we're likely to have a much better vision of what the DELTA system would cost*".

A week later

The Innovation-Purchaser learns from the Process Manager that he is going the next day to MixMat and the test center to perform a second series of tests. This was rushed at the request of the Program Development department. They want tests to be done without the H insert and others to be done with less thickness. On the other hand, the changes on the mold have not been made. So, there will be no production of a part with a good appearance. The consequence is that it will have to be repaid later for new tests. When the Innovation-Purchaser asks him if he will go to the test center with the Materials Experts, he answers that he does not see why; "*it is not needed*".

During this test session, Mario asks to use a transparent material. It blocks the mold – without breaking it. Also, for him, he will receive an invoice because "*Mercury is careful and says it's not his fault. He is not going to give a gift because he is losing money*". Ten days later, this invoice arrives.

According to Mario, "*they did not go too hard. Maybe because they do not really know what to do next with DELCAR. Nothing has been said about the rest – so they do not know if they have been dropped or not*". The next step would be to grain the mold to make good-looking parts, but "*for the moment there are no more expenses of this type whilst there is no budget, so it could be done later at DELCAR*".

In the 10 days which follow this second test session, there are no exchanges between DELCAR and MixMat, i.e. between Mario and Mercury. The last message from Mercury was about a rumor heard in a trade show: the other division of DELCAR would begin to request a quote from suppliers for the DELTA project's serial mold. He is surprised that he is not in the loop and wants to know how to be on their panel. However, Mario has never heard of this request but only quantified evaluation of the possibility of going into series. He must inquire before contacting Mercury again.

The Innovation-Purchaser, at the coffee machine, asks the two Process Managers what they think of MixMat. For Mario, they are no worse than other mold makers. He has already had bad relationships with mold makers, but would not classify this as a bad one. For Luigi, technically they are up to it. They responded to what was asked of them.

If they had to do it again, both would go with MixMat again as "*they responded to what we asked for*". As for what should then be improved, for them, the relationship should be improved because "*at MixMat, it's difficult, they are rigid... but technically the objectives are achieved, the mold works*".

On the same day, the Project Manager is waiting for a response from VROUM to know if the DELTA project can be proposed to the quotation request. Internally at DELCAR, "*they pushed the classic solution well and the DELTA solution served as a platform to get that one off the ground*". The Quotation department estimates that the cost of the traditional system is lower than that of the DELTA system.

A week later, at the beginning of April, the news is sent by the Project Manager to everyone involved in the DELTA project:

> **From: Zeus**
>
> To: The Project-team, Innovation-Purchasing
>
> The DELTA system will not be retained on VR2020. Nevertheless, VROUM has emphasized the very good technical level of our project. We carry out the project up to milestone M3 (equivalent of "TRL6" at VROUM).
>
> This decision comes from the fact that VROUM cannot converge on the profitability criteria set on the VR2020 projects following the DELTA response...
>
> On the other hand, VROUM seems confident regarding a possible connection to a future niche vehicle or limited volumes...

The following week

Mercury writes to the Innovation-Purchaser, copying his Line Manager, MixMat's Sales Manager, and DELCAR's Process Manager. He requests the payment of an invoice which represents half of the total amount of the order. The Process Manager sends this email to the administrative assistant of the innovation floor who will settle the question with MixMat's accountant – "*There may be a problem with the invoice amount*".

The DELCAR teams simply have to sign a tooling voucher, which will then be sent by MixMat – the problem is quickly solved.

During the following meetings between Innovation-Purchasing and the Materials Experts, when the subject touches upon the DELTA project, they do not fail to recall that the solution chosen for the tool to solve the G difficulty prevents industrial feasibility approval. Dionysus, the Materials Expert specialized in mold design, also recalls that he was the first to indicate this, and that the experts of the other division of the DELCAR group later confirmed it. They also add that DELCAR's "classical" technology experts pointed out that "*this is the first time that a crash test system passes everything the first time... it almost never happens on a conventional system ... it means that the DELTA system is not optimized at all*". They also point out that the α material does not flow enough, which means that there are still many optimizations to be made.

The tool for its part is still at the test center. Mario "*does not move because there is nothing signed saying that the DELTA project has stopped. VROUM said he did not want it, but there are still some tasks to be done by June because we have not done everything yet, especially on the issues of appearance. It should be defined as to make the mold being grained and then manufacture new pieces – in parallel we have tests to finish on the parts already produced. So, as long as there is nothing signed in the orders to stop, we will continue, and we will make the requests we need. After we see when the leaders will take a position*".

Regarding the tool, he would like it to be repatriated, "*the transport is included in MixMat's price but then it will have to be stored somewhere at DELCAR. We can do that, but it will be necessary to rent a crane to move it and it will cost at least 3000 euros and we do not know if it will be possible. So, we continue and wait for contradictory orders*".

End of April 2014

The Innovation-Purchaser talks with Zeus and Mario. "*The DELTA system was not in the strategic plan. And anyway, it would never pass on VROUM because for such important volumes to provide a sporty design, the classic solution is much cheaper... just add two prostheses here and there on the conventional system to give it the sought-them look*".

June 2014

The DELTA project is presented to the monthly Innovation Committee to pass the last gate, which will allow them to "*put the product on the shelf and say that the product is robust and everything – and then it's over*".

Gate M3 is passed. The DELTA project could be presented in an offer to another manufacturer.

3.2.10.1. *Analysis: last episode of the collaboration: the delivery*

The last episode of the cooperation between DELCAR and MixMat on the DELTA project corresponds to the finalization of the tool and to the development of the prototypes that are followed by tests within DELCAR.

This last phase is an extension of the previous one. The management of the relationship does not change. The atmosphere shows a rise in DELCAR's trust in MixMat: the quality of the work done by them is accepted and the automobile manufacturer's deadlines are adhered to. The tests on the innovative product have been successful; the project was able to take a new step in the innovation management system. For DELCAR, the goals of the relationship were fulfilled even though the relationship was difficult (Figure 3.22).

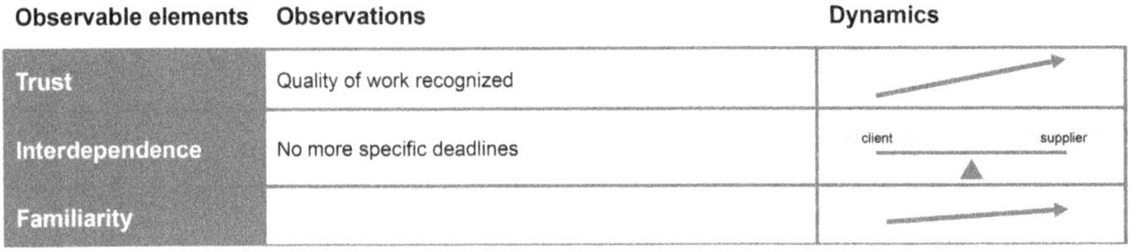

Figure 3.22. *Atmosphere of the relationship during the delivery episode*

On MixMat's side, the determined price has been paid by DELCAR. If MixMat's invoices were not settled by DELCAR in the expected time, the delays did not have any apparent effect on the cooperation other than the call for Innovation-Purchasing to have them settled.[15].

Finally, this relationship has allowed both companies to gain experience: DELCAR has learned about the design of the innovative product and its production process; MixMat has earned a benchmark for making a complex tool for a new application for the automotive industry.

15 Failure to meet payment deadlines in France is a custom, fought by the public authorities both by law (Law of Modernization of the Economy of 2008 and code) and by more positive incentives such as the Charter of the Responsible Supplier Relations. It thus constitutes the first ground for referral to inter-firm mediation and a source of acknowledged tension between actors in the automotive industry [GEN 13]. Also, we could assume that these delays are considered by the supplier as a standard part of business life.

3.3. Feedback on the collaboration – the contribution of the relational approach to the "life cycle" and "teleological" approaches to the dynamics of inter-firm cooperation

This daily story of a collaborative innovation and the analysis of its dynamics are full of lessons to learn, beyond the staff's resilience when facing fluctuations in the environment of the relationship.

3.3.1. *A model for reading the dynamics*

From the point of view of the researcher, it makes it possible to show the relevance of the approach of the dynamics of the client-supplier Open Innovation by the atmosphere of the relationship, the governance of the client–supplier couple, the maturity of the innovation project and the performance of the relationship. The life of a cooperation of innovation can be analyzed by the prism of these different dynamics and their links – by focusing on the inter-firm relationship while taking a step back from the adventures inherent in any human project.

By synthesizing these 20 months through the analysis of the dynamics of the Open Innovation relationship, the following vision of the evolution of cooperation can be proposed (Figure 3.23).

At each episode of the relationship, there correspond a specific governance, a distinct atmosphere and a growing maturation of the innovation project. All are intimately connected.

These developments do not follow a long calm river where the relationship would gradually move from a phase of discovery of the other to more interactions, more cooperation, more trust and so on, until reaching a plateau of harmonious stability. The theoretical approach of the "life cycle" of inter-firm relations described in Chapter 1 is here in default. It seems to be "best practice"[16] rather than current practice.

16 Which is in fact the case since we have come across some in the previous chapter, and a considerable number; but this does not reflect that which has been lived, that lives or will be lived by the majority of managers of client–supplier Open Innovation.

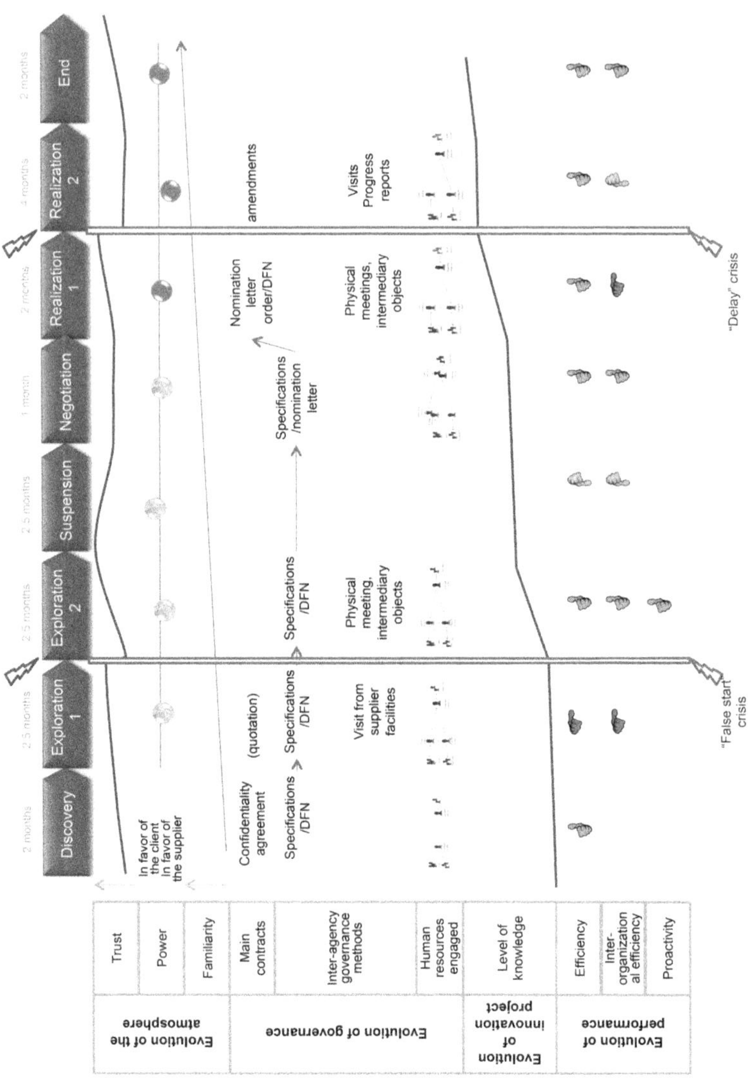

Figure 3.23. *Evolution of the collaboration. For a color version of this figure, see www.iste.co.uk/servajean/dynamics.zip*

A "teleological" approach seems more appropriate to understand the dynamics of client–supplier Open Innovation, an approach in which the relationship evolves, from its beginning, through a continuous cycle of achievements, evaluations and adjustments. But a teleological approach that feeds on the first phases of discovery of the life cycle approach to diverge into trajectories as diverse as the contexts of firms involved in Open Innovation (Figure 3.24).

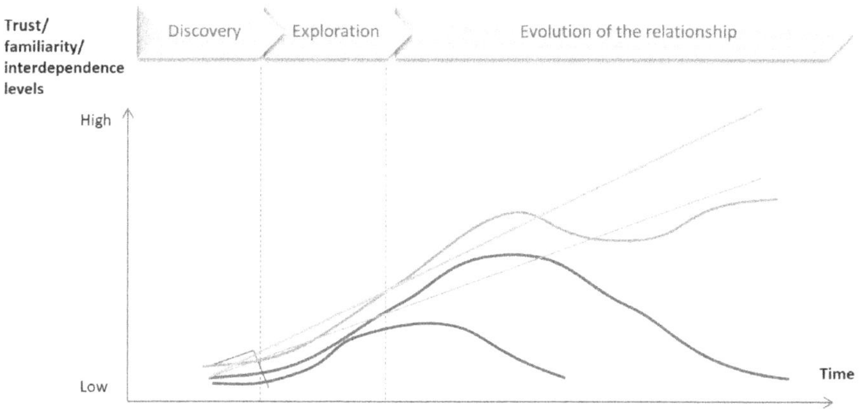

Figure 3.24. *Trajectories of the client–supplier Open Innovation relationship. For a color version of this figure, see www.iste.co.uk/servajean/dynamics.zip*

This approach involves looking at the relationship from the very first interactions and not from the "classic" teleological approach as soon as the initial conditions (definition of the respective expectations, contract, etc.) are established.

The story retold here shows how these first interactions contribute to the exploration of governance modalities in conjunction with the exploration of objectives for the innovation project itself, but also to the design and exploration of innovative solutions.

While the supplier is not recognized as being involved in the project (the only contract that binds it is a confidentiality agreement), the relationship is engaged. The effects of time can already be felt on the atmosphere as well as on governance. Learning is born from this relationship. Decisions are made on the trajectory of innovation and how to manage the relationship.

Expectations like the interface structure are not initially defined, but are constructed during this "silent" involvement of the supplier with the client.

Interest and temptation of the supplier's involvement before their involvement

The "silent" involvement of the supplier in the project from the first interactions, i.e. before being selected, contributes to progressing an innovation project as it emerges, and it is still difficult to specify the target to be reached.

There is then a co-construction of the object of cooperation. The atmosphere specific to the relationship is drawn there. How should governance adjustments be made internally as well as between companies?

For the client, learning from this supplier can be important. For him, the temptation to adopt behaviors of "looting of ideas" therefore comes; behavior that remained at the temptation stage in the DELTA project.

It is indeed possible for a client firm, through quotations and meetings with suppliers, to progress its innovation projects by limiting its financial commitments[17].

Any legal or ethical considerations aside, this behavior is extra risky. A first risk is that its repetition and the multiplication of disappointed potential suppliers lead to a reputation which is detrimental to the future development of Open Innovation. A second risk is financial: through the multiplication of consultations, the cost of the effort can exceed the gains obtained. A third risk is the multiplication of opportunities, leading to a loss of confidential information.

It is from these initial conditions resulting from the first interactions that launches the trajectory of the Open Innovation relationship, continuous cycle of achievements on the project, assessments and adjustments of governance, with the possibility, at any time, to enter a virtuous or vicious cycle, and get out of it.

17 We can find a reciprocal position on the part of supplier companies that come to participate in collaborative projects (funded by the state or Europe) not for innovation but for relational purposes: they allow them to stay in contact with the R&D teams of their clients and maintain/develop their familiarity with them.

For the practice, this vision allows managers to not remain locked in a constant search for "best practice", but rather to seek to constantly adapt to the context, fluctuating in essence but also by necessity – for the required innovation. It is a question of knowing how to transform one's practices, internally and through external interactions, to seek an optimum, a balance – and to do it throughout the time that the cooperation lasts.

3.3.2. *The trust–control link in Open Innovation*

When Open Innovation is mentioned, the first images that come to mind are those of an open world where sharing and trust reign. Then open world goes hand in hand with a world without barriers.

In the transition to practice, reality reiterates the order. Trust cannot rule. Opening does not mean a total absence of barriers. On the contrary, the presence of barriers, safeguards as limits, nurtures trust.

In this section, we have seen how the atmosphere and the governance of Open Innovation are linked. We have also seen how much more trust and control mechanisms and contracts are linked.

By setting up contractual arrangements and then adapting them to changing contexts, each increases the predictability of the behavior of the other despite and through these changes, and therefore its trust.

As such, the contractual framework that frames client–supplier Open Innovation is more than a contract, a document with many clauses. Here, it also consisted of the specifications and the numerical definition of the piece to be made. And this contractual device was initially a coordination mechanism to agree on the target and the expectations of each other; and then stayed.

The contractual system as a whole was built during the relationship both internally at the client side, through the mobilization of the various experts and the increase in its precision, and externally through the exchanges with the supplier (and other pre-qualified suppliers). It made it possible to fix and make explicit the expectations and know-how of each other and their constraints, the use of these multiple mechanisms.

Through it, the object of cooperation has been co-constructed, and not just negotiated. The contractual system is a factor of cohesion *a priori* of the signature of a mutual commitment (the contract concluded) by these exchanges which make it possible to develop a common vision.

It is a factor of cohesion *a posteriori* by allowing the resolution of conflicts within the relationship, either while referring to it or by amending it.

It also reinforces the interdependence: the contractual commitment is to fix the obligations of each one, the endorsements come to specify them and to engage more ahead of the specificity of the investments of each one in the relation.

Moreover, and at any point in the relationship, when trust is undermined, following a rumor about a protagonist or a crisis in the relationship, control mechanisms are put in place; for example, evaluating the financial strength of MixMat or submitting detailed progress reports.

These control mechanisms allow the transmission and reception of information likely to (re)ensure the achievement of project objectives. They help to develop common standards for understanding the project and its challenges – on the project's progress and the possibility of reaching the target even more than on the target itself. Trust is revived.

Similarly, the involvement of Innovation-Purchasing in the project is a mechanism that is much related to the atmosphere of the relationship and the need to safeguard the interests of the client as part of the collaboration.

In this section, Innovation-Purchasing, a new function of Open Innovation [SER 18b], is mobilized at times of low trust: during the discovery and exploration episodes that lead to negotiation and contracting, then during the "delay" crisis.

This function is mobilized by the Innovation team when the need for formal control is felt, as they maintain a relationship with the supplier based on the recognition of the technical capabilities of everyone around the object of cooperation [BRA 13]. It is mobilized by the supplier when it comes to solving difficulties impacting the commercial relationship: getting bills paid and finding a solution to a delay.

This is a purchasing function. It accompanies innovation projects in the exploration phase in the search for suppliers and the construction of appropriate contracts. It uses purchasing tools, specifications, evaluation and classification of suppliers and so on.

But this is not a "classic" purchasing function. It is complementary to this one. It adapts its tools to the contexts of the Open Innovation, softens them according to the changes. It thus helps to build an objectified vision of the object of cooperation, internally and externally, and helps to forge the most appropriate contractual system.

It can also mobilize the tools and attitudes of Purchasing – especially in the negotiation phases. By distributing the roles of "good cop/bad cop" against the supplier with Innovation, Innovation-Purchasing contributes to increasing trust.

The "good" role is assigned to Innovation staff who develop a good relationship with the R&D staff of the supplier. While the negotiation is in full swing, tensions remain crystallized on Purchasing; the collaboration continues around innovation. Once the problem is solved, the "bad" role is put aside until next time[18].

And trust starts again on the bases established between those who work on the object of the cooperation.

Functions such as those described are developing, with the development of Open Innovation, but they are still rare. This description of work practices, "ordinary" actions, and the impacts of such a function can facilitate the setting up of similar entities in companies that wish to open their innovation to their suppliers.

3.3.3. *Learning and exchanges in real life*

Throughout the relationship, email has been a very important means for transferring information between companies and, at least, within the client

18 If this approach has the merit of allowing a continuity of the level of trust during a tense period, it is to be guaranteed that the involvement of Innovation-Purchasing throughout the project could possibly have limited certain glitches and the difficulties of relational orders that subsequently arose. But if these still appeared, would this division of roles, and its positive impact on the atmosphere and the performance of the relationship, have been possible?

firm. However, it appears that the production of new knowledge was most noticeable when transfers were "in real life".

Within the client, the most progress in terms of understanding the context, understanding the other and building the project target was achieved in three places.

The first is the "work island" on the innovation floor. This workspace brought together in one place the entire Innovation team. It was the place of the weekly team meeting. The one where everybody met before the meetings, where the meetings of crisis were organized, where the prototypes and the digital models were manipulated and where the negotiations were prepared.

The second place was that of the coffee machine closest to the innovation floor. This is the place where briefings and debriefings of internal meetings and visits to the supplier took place. It is also the place where the feelings of each other were collected on the inter-firm relationship. The third is simply the nearest outdoor ashtray, an annex of the coffee machine.

Between the two firms, the most important learnings, those that have matured the purpose of the cooperation, the innovation project, have occurred on the occasion of two meetings "in real life". On this occasion, the staff of both organizations were gathered in the same room.

Beyond specifications and numerical definitions, information was exchanged through multiple intermediate objects. The client used models and samples to present to the supplier what is expected. Both of them exchanged around three-dimensional representations, around two-dimensional animated representations and around diagrams to expose the technical stakes encountered and to discuss possible solutions.

The impact of these mechanisms, direct contacts, physical objects and manipulable virtual objects on the innovation project was equivalent, during these meetings, before and after the supplier was contractually integrated into the project (order contract passed).

We thus find the results of work on the transfer of knowledge through direct physical exchanges involving intermediate objects [BRE 13, GAR 96, LAK 06a, LED 14], while noting the lack of influence of the atmosphere of the relationship on this learning. Whether the atmosphere is good or not so good, the quality of the exchanges in terms of mutual learning has been the same when the exchanges were made live and with intermediate objects.

3.3.4. *Sources and dynamics of the performance*

Through the analysis of the DELTA project, the links between the atmosphere and the performance are mainly identified through the variations of efficiency during the relationship. These variations in efficiency are seen on the client side, in the light of the expression of the difficulties of inter-firm coordination.

Efficiency appears low when trust is also high and supplier dependence is high. It is higher in episodes where trust is highest, and interdependence is low; either at the beginning of the relationship when there are no formal commitments and there is a strong willingness to work together, or at the end of the project when the main commitment has already been kept, i.e. the main objectives of the project are kept.

The links between governance and performance are more difficult to reveal. They are done through the identification of the governance mechanisms that led to a non-performance.

Regarding the effectiveness of the relationship, it appears that the achievement of the goals of the relationship is at the rendezvous at each episode of the relationship, with three exceptions.

These three exceptions are the two moments of strong tension, namely the "false start" of December 2012 and the "urgent delay" of September 2013, and the episode of suspension of the relationship of spring 2013. The first two have their origin in the dysfunction of the internal coordination with the client, then with the supplier. The latter is the result of the timing of the innovation project – the passage of a project milestone delayed until further technical and commercial safeguards were obtained. So, it was the smooth running of the internal project management procedure with the client which had a negative impact when it was implemented – even though the impact on the soundness of the project, subsequent to the collaboration, has been positive in the long run.

The efficiency of the relationship varies with each episode mainly according to the atmosphere of the relationship.

Depending on whether the supplier's role in the innovation project appears critical or not to the client, the mechanism of the single interlocutor and the "batch" information transmission from the supplier appears more or less efficient.

Efficiency was also negatively impacted by two recurrent sources of inter-organizational glitches. During the silent involvement of the supplier, i.e. before his formal commitment, there was almost constantly a lack of correspondence between the client's requests (his specifications and the requested response formats) and the supplier's responses to those requests (the form and content of the quote).

Then, as soon as the supplier has been formally involved, when there is a signed contract, the Innovation-Purchaser[19] was excluded from the exchanges by the client's project team and by the supplier – outside the times when his intervention was requested.

Proactivity was only noticed during the exploration episode of potential collaboration during the first winter – and only from the supplier. This one brought to its potential client solutions and ways of improvement without being formally solicited. This proactivity has helped to increase client's trust in the vendor's ability to meet their needs. It has also contributed to its being recognized as better than other potential suppliers.

The question that arises then is to what extent does this proactivity derive from the mechanisms used at the time (face-to-face meeting and intermediate objects) and/or from the supplier's commercial approach (efforts made due to the importance of the client, and the opening of the automotive market on this technology, for his firm).

In conclusion, the goals of the relationship as a whole have been achieved.

---

19 The non-integration of experts in certain episodes of the relationship appears to us as other potential sources of glitches just as their integration would also have been a source of higher coordination costs. Their impact on the relationship – both positive and negative – is more difficult to identify.

An evolutionary mechanism of central coordination of the relationship has made it possible to align the objectives of the couple and of each firm: the specifications that included the numerical definition of the innovative system, this specification being first of all the selection mechanism of the supplier up to being a contractual mechanism, reference of the object of cooperation. In addition, internal project management and relationship management mechanisms played their role in safeguarding the interests of each organization by limiting drifts.

For the client, this included, for example, the verification of technological feasibility when passing through gates, from the participation of Innovation-Purchasing to the selection of the supplier and then to the negotiations, or the integration of room for maneuverering in the planning and contractual agreement. For the supplier, this included, for example, verification by the Purchasing entity of the project budget, or the management of technical requests through a single point of contact. These internal safeguards have strongly contributed to the overall performance of the collaboration, even though they occasionally have had a negative impact on the relationship.

Conclusion

The objective of the research of which the results are presented in this book was to study the dynamics of client–supplier Open Innovation. The aim was to understand how the performance of the client–supplier relationship is built and evolves in the context of the shifting and uncertain environment of innovation.

It started from a need identified as a practitioner: the difficulties that companies face, of any size, any sector and whatever their position in a value network, to set up and then maintain innovation relationships from the exploration of new methods all the way to their exploitation.

This need was also in line with the political and economic actors' reading of the need to make client–supplier relationships partnership-based and to stimulate collective intelligence in order to strengthen the competitiveness of business ecosystems [GAL 12, RET 11, VOL 10].

Previous academic works shed light on this subject, or part of it. However, only a few allow us to understand its complexity. Moreover, in order to contribute to the answer to these needs, our research path carried out within the framework of a doctoral thesis [SER 15] brought us to confront theories and perceptions of the reality.

It allowed us to reach a level of understanding of the issues that were difficult for the practitioner to access when they were subjected to the pressure of the firm's time. Following this path, we offer practitioners and theorists a better understanding of the complexity of the dynamics of Open Innovation between a client and a supplier.

It is possible through the conceptual model of vertical innovation collaboration that we propose. This model consists of a new way of understanding client–supplier Open Innovation, to read and control its dynamics and its performance – by focusing more on the firms as a couple than on one of the two firms. Through this and the analyses drawn from the observation of the multiple cases studied, we can better understand the management mechanisms of Open Innovation. Especially for some cases, we understand more how, depending on the conditions surrounding the client–supplier relationship in innovation, they can contribute to the construction of the performance.

In the following sections, we present in more detail these two contributions to the knowledge of client–supplier Open Innovation.

A conceptual model of vertical innovation collaboration

The research results presented here have enabled the construction and validation of a conceptual model (Figure C.1), the elements of which have been presented in this section.

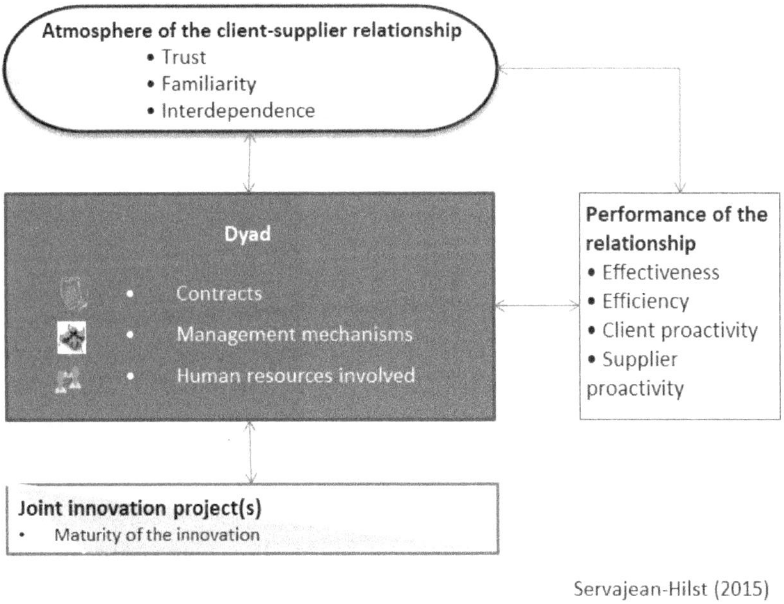

Figure C.1. *A conceptual model of vertical Open Innovation collaboration*

This model describes the dynamics of the interactions of client–supplier Open Innovation with regard to the performance of the relationship. It consists of three major performance components:

– the governance of the couple made up of the different mechanisms that govern the interactions within the relationship: the contracts, the management mechanisms and the human resources involved in the collaboration;

– the atmosphere of the relationship, inter-firm trust, client–supplier interdependence and familiarity;

– the joint innovation project, through its stages of maturity, from the moment when the interactions begin and its maturation over time.

Our model is based on the interaction model of the IMP group, which describes interactions within a client–supplier relationship. It comes to adapt this descriptive model to a specific object which is the situation of innovation between a client and a supplier. Our conceptual model also complements that of the IMP group [HÅK 82] by its explanatory nature. It allows us to examine and understand how a client and a supplier manage their collaboration during innovation projects that they lead together, by looking at the client–supplier relationship.

A measure of the performance of the Open Innovation relationship

Our work allowed us to propose and validate a measure of the performance of the client–supplier relationship in innovation. For this purpose, we relied on the work of Le Dain, Calvi and Cheriti [LED 11] to measure the performance of a supplier. We adapted them for a unit of analysis, that is, the client–supplier couple in the context of innovation.

We then proposed a scale of performance measurement according to four variables: (1) the effectiveness and (2) the efficiency of the relationship, (3) the proactivity of the supplier and (4) the proactivity of the client – these last two represent an outperformance of the relationship.

We supplemented it by adding to the measure of efficiency the measure of non-efficiency by relying on the concept of glitch [HOP 99].

Moreover, the clash between theories and practices during this research led us not to consider trust as a component of the performance of the relationship but of its atmosphere. If they are related, trust and performance of the relationship are two very specific concepts.

We observed that there could be high levels of performance without trust, but that without a high level of trust, performance could never reach the highest levels. A high level of trust is a necessary but not sufficient condition to achieve the highest performance, with high performance resulting in increased trust.

Through this contribution, we participate in understanding the elements that constitute a relationship of vertical innovation collaboration and influence its performance over time. By proposing to understand the dynamics of Open Innovation through the adoption of the unit of analysis "the couple" – and not a single firm – and through the measurement of the performance of the relationship, our model also allows us to move outside of the market logic that prevails in the academic currents of Open Innovation, such as the involvement of suppliers in innovation projects. We leave the dominant transactional approach to enter a relational logic.

This leads us to also provide a complementary view of the experience preceding and following the innovation project, especially by taking into account the atmosphere of the relationship. Although the work on suppliers' involvement in their clients' innovation projects has so far largely emphasized the importance of this dimension, only a few integrated it into their analysis grids. To the cognitive dimension relative to innovation, we add the cognitive dimension relative to the relationship: the learning is done on both the innovation and the relationship side.

In-depth knowledge of the management of client–supplier innovation collaboration

The combination of several forms of research and several academic currents has allowed us to produce a wide variety of results to better manage the dynamics of client–supplier Open Innovation. In addition to the conceptual model that we have just presented, we have been able to offer ways to better manage this type of collaboration. Through a questionnaire survey, the results of which are presented in Chapter 2, we identified four main management methods adopted by each client–supplier couple interacting on Open Innovation projects. Then, we identified which of these management methods are the most successful according to the atmosphere of the relationship and the maturity of the project at the time of the supplier's involvement.

We also observed how the partnership forms were likely to favor the best performance – whether the partnership is reflected in the quality of the inter-firm atmosphere or in the management. In this regard, we observed the importance of the commitment to sharing the value on an innovation project in the exploitation phase, beyond the sharing of intellectual property rights. It was then present at the contractual level, with the resale rights in the event of the project's success, as well as at the level of the functions strongly implicated in the relationship, not only those related to innovation but also those which later will manage its exploitation.

This means the interest that there may be, when negotiating a contract on a collaborative innovation project, in examining the commitments that the client and the supplier can make to each other, in the case of success of the innovation project, in order to jointly exploit the results of this project. This therefore implies not stopping the reflection on the intellectual property and the distribution of the related rights but extending it to the *supply chain*. Co-designing, and contracting, the joint innovation business model in its entirety makes it possible to set the terms for the co-creation of value and also the co-capture of value, with the effect of reinforcing and securing this capture, thanks to the relational income generated by this co-design and joint commitment.

Then, through the study of a client–supplier Open Innovation project presented in Chapter 3, we completed our macro vision with a micro vision of the management of collaboration and its dynamics.

This allowed us to identify a new "Open Innovation Champion": the Innovation-Purchaser. We could see their pivotal role in the client–supplier Open Innovation and the use they made of the purchasing tools to adapt the tools to the context of innovation.

Innovation-Purchasing: pivotal function of Open Innovation

By completing the "Open Innovation Champions" panel [CHE 06a] with this Purchasing function dedicated to Open Innovation, we have highlighted its pivotal role in helping internal and inter-firm harmony.

Complementary to the Project Manager, a pivotal role of the innovation project, the Innovation–Purchasing function is the pivotal part of the innovation relationship for its firm.

In the negotiation phases, it helps to preserve the room for maneuver of each part of the relationship by adapting the existing tools to the specificities of the innovation project. It also plays a role of translator and gatekeeper – with the supplier and also between the various entities in-house, notably the Purchasing function and the Innovation function.

Then, it guarantees the successful implementation of the relationship's objectives with this role of "trouble-shooter", which is facilitated by a discontinuous involvement in the relationship and is partially misaligned from R&D to the supplier [BRA 13, MEL 15]. When the atmosphere of the relationship deteriorates, its intervention as a control agency realigns organizations and regains lost trust.

Its position as an influential outsider, a stakeholder of the two systems (Purchasing and Innovation often being presented as opposing terms), allows it to respond both to the need for Innovation to create tailor-made models and, if necessary, to Purchasing to follow the traditional processes of supplier risk management.

We also highlighted the evolving nature of Open Innovation governance. If there are macroscopic governance modalities, and optimal modalities, this optimum is also at a more microscopic level through a composition and a permanent reconstructing of the elements of these modalities in the context of the relationship.

To this end, we observed the sometimes very rapid variations in the different elements of the couple governance and their repercussions on each other, with respect to each other – especially all those who make it possible to regain trust that has deteriorated. We also observed the hybrid nature of certain management mechanisms, and their variable impacts on atmosphere and performance. The most emblematic in our case is the specifications which have been mechanisms for exchange of information between and within firms, control systems and contractual mechanisms.

In addition, we identified the importance and relative autonomy of information exchange mechanisms “in real life” with regard to the atmosphere of the relationship. Whatever the atmosphere of the relationship, intermediate objects and physical spaces have revealed their central place in the learning capacity between the client and the supplier.

Finally, we highlighted the complementary nature of trust and control mechanisms in an Open Innovation context, which goes against the majority of the literature on the subject. Whether by a quantitative or qualitative approach, we highlighted the same phenomenon: trust does not exclude control and control can foster trust, from the point of view of both the client and the supplier.

Through the DELTA case, we observed how, when trust declined, the activation of control mechanisms made it possible to regain lost confidence, or even to increase it. Through the questionnaire survey, we identified that important contractual and relational control mechanisms were not antithetical to a high level of trust, and that the absence of the same mechanisms favored atmospheres with the lowest trust levels.

Thus, we contribute to the academic debate on the complementary or substituting nature of the control–trust connection [DEK 08]. Following Brattström and Richtnér [BRA 13], we consider that control and trust are two fundamentally different concepts. In addition, it is possible to consider a coexistence: one is not exclusive of the other.

By distinguishing, on the basis of our conceptual model, inter-firm trust and control mechanisms, we showed that high levels of trust and formal control can go hand in hand. We emphasized it both from a dynamic point of view and in an Open Innovation context.

First, for trust to be established in a business-to-business relationship, a minimum level of control is required, for example via a contract or a risk of loss of reputation in the event of opportunistic behavior. Second, control mechanisms can counteract a decline in trust – they are mechanisms for reassessing the relationship that lead to mutual adjustments that then feed into the collaboration.

Trust and control have a mutually beneficial, symbiotic relationship for client–supplier Open Innovation.

References

[ADN 17] ADNER R., "Ecosystem as structure: An actionable construct for strategy", *Journal of Management*, vol. 43, no. 1, pp. 39–58, 2017.

[AHU 01] AHUJA G., KATILA R., "Technological acquisitions and the innovation performance of acquiring firms: A longitudinal study", *Strategic Management Journal*, vol. 22, pp. 197–220, 2001.

[AND 84] ANDERSON J.C., NARUS J.A., "A model of the distributor's perspective of distributor-manufacturer working relationships", *The Journal of Marketing*, vol. 48, pp. 62–74, 1984.

[ARR 71] ARROW K.J., *Essays in the Theory of Risk-bearing*, Markham Publishing Firm, Chicago, 1971.

[ATU 95] ATUAHENE-GIMA K., "Involving organizational buyers in new product development", *Industrial Marketing Management*, vol. 24, pp. 215–226, 1995.

[BAN 11] BANOUN A., DUFOUR L., *L'opportunisme : Une approche pluridisciplinaire*, Hermes-Lavoisier, 2011.

[BAR 38] BARNARD C.I., *The Functions of the Executive*, Harvard University Press, 1938.

[BAR 07] BARNES B.R., NAUDÉ P., MICHELL P., "Perceptual gaps and similarities in buyer–seller dyadic relationships", *Industrial Marketing Management*, vol. 36, pp. 662–675, 2007.

[BEN 04] BEN MAHMOUD-JOUINI S., CALVI R., "Les coopérations interentreprises dans les projets de développement", in GAREL G., GIARD V., MIDLER C. *et al.* (eds), *Faire de la recherche en management de projet*, Vuibert, 2004.

[BEN 07] BEN MAHMOUD-JOUINI S., CHARUE-DUBOC F., FOURCADE F., "Favoriser l'innovation radicale dans une entreprise multidivisionnelle : Extension du modèle ambidextre à partir de l'analyse d'un cas", *Finance contrôle stratégie*, vol. 10, pp. 5–41, 2007.

[BEN 18] BEN MAHMOUD-JOUINI S., CHARUE-DUBOC F., "Construction of relations with distant suppliers on the cognitive and relational dimensions to co-explore discontinuous innovations", *Innovations*, no. 55, pp. 61–87, 2018.

[BER 83] BERRY M., Une technologie invisible – L'impact des instruments de gestion sur l'évolution des systèmes humains, 1983.

[BEW 07] BEW R., "The new customer of choice imperative: Ensuring supply availability, productivity gains, and supplier innovation, *92nd Annual International ISM Conference*, 2007.

[BID 98] BIDAULT F., DESPRES C., BUTLER C., "New product development and early supplier involvement (ESI): The drivers of ESI adoption", *International Journal of Technology Management*, vol. 15, pp. 49–69, 1998.

[BID 99] BIDAULT F., DESPRES C., BUTLER C., *Leveraged Innovation: Unlocking the Innovation Potential of Strategic Supply*, St. Martin's Press, 1999.

[BLA 64] BLAU P.M., *Exchange and Power in Social Life*, Transaction Publishers, 1964.

[BOY 95] BOYLE B.A., DWYER F.R., "Power, bureaucracy, influence, and performance: Their relationships in industrial distribution channels", *Journal of Business Research*, vol. 32, pp. 189–200, 1995.

[BRA 13] BRATTSTRÖM A., RICHTNÉR A., "Good cop–bad cop: Trust, control, and the lure of integration", *Journal of Product Innovation Management*, vol. 31, pp. 584–598, 2013.

[BRE 13] BREITE R., KOSKINEN K.U., "Boundary objects as a potential co-ordination mechanism in knowledge communication of the supply chain", *22nd IPSERA Annual Conference*, pp. 640–652, 2013.

[CAL 00] CALVI R., "Le rôle des Services Achats dans le développement des produits nouveaux : Une approche organisationnelle", *Finance, contrôle et stratégie*, vol. 3, pp. 31–55, 2000.

[CAL 13] CALVI R., LE DAIN M.-A., "Partager la conception avec ses fournisseurs", *Revue excellence HA*, vol. 1, pp. 27–37, 2013.

[CAN 99] CANNON J.P., PERREAULT W.D.J., "Buyer-seller relationships in business markets", *Journal of Marketing Research*, vol. 36, pp. 439–460, 1999.

[CAP 12] CAPPELLI G., *Réinventer notre industrie – Le rôle essentiel des achats*, Le Club Rodin, 2012.

[CAR 10] CARRICANO M., POUJOL F., BERTRANDIAS L., *Analyse de données avec SPSS*, Pearson, Paris, p. 235, 2010.

[CAV 12] CAVEROT G., PME impliquées dans les projets d'open innovation : Introduction d'un "technological gatekeeper" pour favoriser la création de valeur et améliorer la performance, IGR-IAE de Rennes, Rennes, 2012.

[CHE 03] CHESBROUGH H., *Open Innovation: The New Imperative for Creating and Profiting from Technology*, Harvard Business Press, Boston, 2003.

[CHE 06a] CHESBROUGH H., CROWTHER A.K., "Beyond high tech: Early adopters of open innovation in other industries", *R & D Management*, vol. 36, pp. 229–236, 2006.

[CHE 06b] CHESBROUGH H., VANHAVERBEKE W., WEST J., *Open Innovation: Researching a New Paradigm*, Oxford University Press, 2006.

[CHE 11] CHERITI S., Conception collaborative : Propositions pour construire et piloter des relations performantes avec les fournisseurs, Institut national polytechnique de Grenoble, 2011.

[CLA 91] CLARK K.B., FUJIMOTO T., *Product Development Performance: Strategy, Organization, and Management in the World Auto Industry*, vol. 432, Harvard Business Press, Boston, 1991.

[COH 90] COHEN W.M., LEVINTHAL D.A., "Absorptive capacity: A new perspective on learning and innovation", *Administrative Science Quarterly*, vol. 35, pp. 128–152, 1990.

[COO 90] COOPER R.G., "Stage-gate systems: A new tool for managing new products", *Business Horizons*, vol. 33, pp. 44–54, 1990.

[COU 07] COUSINS P.D., LAWSON B., "The effect of socialization mechanisms and performance measurement on supplier integration in new product development", *British Journal of Management*, vol. 18, pp. 311–326, 2007.

[CRO 01] CROOM S.R., "The dyadic capabilities concept: Examining the processes of key supplier involvement in collaborative product development", *European Journal of Purchasing and Supply Management*, vol. 7, pp. 29–37, 2001.

[CRO 10] CROPPER S., PALMER I., "Change, dynamics and temporality in inter-organizational relationships", in CROPPER S., HUXHAM C., EBERS M. *et al.* (eds), *The Oxford Handbook of Inter-Organizational Relations*, Oxford University Press, 2010.

[DAS 98] DAS T.K., TENG B.-S., "Between trust and control: Developing confidence in partner cooperation in alliances", *The Academy of Management Review*, vol. 23, pp. 491–512, 1998.

[DAS 00] DAS T.K., TENG B.-S., "A resource-based theory of strategic alliances", *Journal of Management*, vol. 26, pp. 31–61, 2000.

[DAS 02] DAS T.K., TENG B.-S., "The dynamics of alliance conditions in the alliance development process", *Journal of Management Studies*, vol. 39, pp. 725–746, 2002.

[DAY 10] DAYAN M., DI BENEDETTO C.A., "The impact of structural and contextual factors on trust formation in product development teams", *Industrial Marketing Management*, vol. 39, pp. 691–703, 2010.

[DEE 03] DEEDS D.L., ROTHAERMEL F.T., "Honeymoons and liabilities: The relationship between age and performance in research and development alliances", *Journal of Product Innovation Management*, vol. 20, pp. 468–484, 2003.

[DEK 04] DEKKER H.C., "Control of inter-organizational relationships: Evidence on appropriation concerns and coordination requirements", *Accounting, Organizations and Society*, vol. 29, pp. 27–49, 2004.

[DEK 08] DEKKER H.C., "Partner selection and governance design in interfirm relationships", *Accounting, Organizations and Society*, vol. 33, pp. 915–941, 2008.

[DON 05] DONADA C., NOGATCHEWSKY G., "Vingt ans de recherches empiriques en marketing sur la performance des relations client-fournisseur", *Recherche et Application en Marketing*, vol. 20, pp. 71–96, 2005.

[DON 07] DONADA C., NOGATCHEWSKY G., "La confiance dans les relations interentreprises. Une revue des recherches quantitative", *Revue française de gestion*, vol. 33, pp. 111–124, 2007.

[DON 08] DONADA C., NOGATCHEWSKY G., "Partenariat, vassalité, marché et seigneurie : 4 configurations de contrôle client-fournisseur", *Comptabilité – Contrôle – Audit*, vol. 14, pp. 145–168, 2008.

[DOW 92] DOWLATSHAHI S., "Purchasing's role in a concurrent engineering environment", *International Journal of Purchasing and Materials Management*, vol. 28, p. 21, 1992.

[DOZ 87] DOZ Y.L., "Technology partnerships between larger and smaller firms: Some critical issues", *International Studies of Management and Organization*, vol. 17, pp. 31–57, 1987.

[DOZ 96] DOZ Y.L., "The evolution of cooperation in strategic alliances: Initial conditions or learning processes?", *Strategic Management Journal*, vol. 17, pp. 55–83, 1996.

[DOZ 99] DOZ Y.L., OLK P.M., RING P.S. *et al.*, "Formation processes of R & D consortia: Which path to take?", *Where does it lead? INSEAD*, 1999.

[DUM 09] DUMEZ H., "Qu'est-ce qu'un dispositif? Agamben, Foucault et Irénée de Lyon dans leurs rapports avec la gestion", *Le Libellio d'Aegis*, vol. 5, pp. 34–39, 2009.

[DUM 13] DUMEZ H., *Méthodologie de la recherche qualitative; Les 10 questions clés de la démarche compréhensive*, Vuibert, Paris, 2013.

[DUS 95a] DUSSAUGE P., GARRETTE B., *Les stratégies d'alliance*, Eyrolles, 1995.

[DUS 95b] DUSSAUGE P., GARRETTE B., "Determinants of success in international strategic alliances: Evidence from the global aerospace industry", *Journal of International Business Studies*, vol. 26, pp. 505–530, 1995.

[DWY 87] DWYER F.R., SCHURR P.H., OH S., "Developing buyer–seller relationships", *Journal of Marketing*, vol. 51, pp. 11–27, 1987.

[DYE 96] DYER J.H., "Specialized supplier networks as a source of competitive advantage: Evidence from the auto industry", *Strategic Management Journal*, vol. 17, pp. 271–291, 1996.

[DYE 98] DYER J.H., SINGH H., "The relational view: Cooperative strategy and sources of interorganizational competitive advantage", *The Academy of Management Review*, vol. 23, pp. 660–679, 1998.

[DYE 02] DYER J.H., OUCHI W.G., "Japanese-style partnerships: Giving companies a competitive edge", *Managing Innovation and Change*, vol. 203, 2002.

[EIS 95] EISENHARDT K.M., TABRIZI B.N., "Accelerating adaptive processes: Product innovation in the global computer industry", *Administrative Science Quarterly*, vol. 40, pp. 84–110, 1995.

[ELL 12] ELLIS S.C., HENKE JR. J.W., KULL T.J., "The effect of buyer behaviors on preferred customer status and access to supplier technological innovation: An empirical study of supplier perceptions", *Industrial Marketing Management*, vol. 41, pp. 1259–1269, 2012.

[EMD 06] EMDEN Z., CALANTONE R.J., DROGE C., "Collaborating for new product development: Selecting the partner with maximum potential to create value", *Journal of Product Innovation Management*, vol. 23, pp. 330–341, 2006.

[ENG 03] ENGWALL M., "No project is an island: Linking projects to history and context", *Research Policy*, vol. 32, pp. 789–808, 2003.

[ENG 14] ENG T.-Y., OZDEMIR S., "International R & D partnerships and intrafirm R & D–marketing–production integration of manufacturing firms in emerging economies", *Industrial Marketing Management*, vol. 43, pp. 32–44, 2014.

[FAB 17] FABBRI J., GLASER A., GAUJARD C. *et al.*, "Espaces collaboratifs d'innovation : Au-delà du phénomène de mode, de quoi parle-t-on ?", *Entreprendre & Innover*, vols 5–7, 2017.

[FAE 08] FAEMS D., JANSSENS M., MADHOK A. *et al.*, "Toward an integrative perspective on alliance governance: Connecting contract design, trust dynamics, and contract application", *The Academy of Management Journal*, vol. 51, pp. 1053–1078, 2008.

[FRE 04] FRECHET M., *Prévenir les conflits dans les partenariats d'innovation*, Vuibert, 2004.

[FUK 95] FUKUYAMA F., *Trust: The Social Virtues and the Creation of Prosperity*, Free Press New York, 1995.

[GAB 87] GABARRO J.J., *The Dynamics of Taking Charge*, Harvard Business Press, 1987.

[GAL 12] GALLOIS L., Pacte pour la compétitivité de l'industrie française, Report, vol. 74, 2012.

[GAR 94] GAREL G., Réduction du temps de conception, concourance et savoirs professionnels : Le cas de l'emboutissage dans les projets automobiles, Ecole polytechnique, 1994.

[GAR 96] GAREL G., "L'entreprise sur un plateau: Un exemple de gestion de projet concourante dans l'industrie automobile", *Cahiers du Centre de recherche en gestion*, vol. 13, pp. 55–69, 1996.

[GAR 09] GARDET E., "Modes de coordination instaurés par le pivot d'un réseau d'innovation : le cas d'un porteur de projet TPE", *Management & Avenir*, no. 26, pp. 33–51, 2009.

[GEN 13] GENET J.P., "Les relations entre constructeurs et founisseurs restent tendues", *L'argus*, available at: http://www.largus.fr/actualite-automobile/les-relations-entre-constructeurs-et-fournisseurs-restent-tendues-3201373.html, 2013.

[GLA 09] GLASER B.G., STRAUSS A.L., *The Discovery of Grounded Theory: Strategies for Qualitative Research*, Transaction Publishers, 2009.

[GOR 87] GORB P., DUMAS A., "Silent design", *Design Studies*, vol. 8, pp. 150–156, 1987.

[GRA 96] GRANT R., "Toward a knowledge-based theory of the firm", *Strategic Management Journal*, vol. 17, pp. 109–122, 1996.

[GUL 95] GULATI R., "Does familiarity breed trust? The implications of repeated ties for contractual choice in alliances", *The Academy of Management Journal*, vol. 38, pp. 85–112, 1995.

[GUL 98] GULATI R., SINGH H., "The architecture of cooperation: Managing coordination costs and appropriation concerns in strategic alliances", *Administrative Science Quarterly*, pp. 781–814, 1998.

[GUL 05] GULATI R., LAWRENCE P.R., PURANAM P., "Adaptation in vertical relationships: Beyond incentive conflict", *Strategic Management Journal*, vol. 26, pp. 415–440, 2005.

[GUN 95] GUNDLACH G.T., ACHROL R.S., MENTZER J.T., "The structure of commitment in exchange", *Journal of Marketing*, vol. 59, pp. 78–92, 1995.

[HÅK 82] HÅKANSSON H., *IMP Project Group International Marketing and Purchasing of Industrial Goods: An Interaction Approach*, Wiley, Chichester, 1982.

[HÅK 89] HÅKANSSON H., SNEHOTA I., "No business is an island: The network concept of business strategy", *Scandinavian Journal of Management*, vol. 5, pp. 187–200, 1989.

[HÅK 93] HÅKANSON L., "Managing cooperative research and development: Partner selection and contract design", *R & D Management*, vol. 23, pp. 273–285, 1993.

[HÅK 06] HÅKANSSON H., SNEHOTA I., "'No business is an island' 17 years later", *Scandinavian Journal of Management*, vol. 22, pp. 271–274, 2006.

[HAM 12] HAMMERVOLL T., "Managing interaction for learning and value creation in exchange relationships", *Journal of Business Research*, vol. 65, pp. 128–136, 2012.

[HAN 99] HANDFIELD R.B., RAGATZ G.L., PETERSON K. *et al.*, "Involving suppliers in new product development?", *California Management Review*, vol. 42, pp. 59–82, 1999.

[HAN 18] HANNAH D.P., EISENHARDT K.M., "How firms native cooperation and competition in nascent ecosystems", *Strategic Management Journal*, vol. 39, no. 2, pp. 3163–3192, 2018.

[HAT 99] HATCHUEL A., "Connaissances, modèles d'interaction et rationalisations – De la théorie de l'entreprise à l'économie de la connaissance", *Revue d'économie industrielle*, vol. 88, pp. 187–209, 1999.

[HAU 10] HAUSMAN A., JOHNSTON W.J., "The impact of coercive and non-coercive forms of influence on trust, commitment, and compliance in supply chains", *Industrial Marketing Management*, vol. 39, pp. 519–526, 2010.

[HEI 03] HEIDE J.B., "Plural governance in industrial purchasing", *Journal of Marketing*, vol. 67, pp. 18–29, 2003.

[HOE 05] HOEGL M., WAGNER S.M., "Buyer-supplier collaboration in product development projects", *Journal of Management*, vol. 31, pp. 530–548, 2005.

[HOO 99] HOOPES D.G., POSTREL S., "Shared knowledge, 'glitches', and product development performance", *Strategic Management Journal*, vol. 20, pp. 837–865, 1999.

[HUI 11] HUIZINGH E.K.R.E., "Open innovation: State of the art and future perspectives", *Technovation*, vol. 31, pp. 2–9, 2011.

[IMA 85] IMAI K., NONAKA I., TAKEUCHI H., "Managing the new product development process: How Japanese companies learn and unlearn", in HAYES R., CLARK C. (eds), *The Uneasy Alliance: Managing the Productivity-Technology Dilemma*, 1985.

[ING 03] INGHAM M., MOTHE C., "Confiance et apprentissages au sein d'une alliance technologique", *Revue française de gestion*, no. 143, pp. 111–128, 2003.

[JAP 01] JAP S.D., "'Pie Sharing'" in complex collaboration contexts", *Journal of Marketing Research*, vol. 38, pp. 86–99, 2001.

[JOH 08] JOHNSEN T.E., JOHNSEN R.E., LAMMING R.C., "Supply relationship evaluation: The relationship assessment process (RAP) and beyond", *European Management Journal*, vol. 26, pp. 274–287, 2008.

[JOH 09] JOHNSEN T.E., "Supplier involvement in new product development and innovation: Taking stock and looking to the future", *Journal of Purchasing and Supply Management*, vol. 15, pp. 187–197, 2009.

[JOH 11] JOHNSEN T., CALVI R., PHILLIPS W., "Purchasing and supplier involvement in discontinuous innovation: A literature review", *20th Annual IPSERA Conference*, 2011.

[JOU 08] JOURNE B., "Collecter les données par l'observation", in GAVARD-PERRET M.-L., GOTTELAND D., HAON C. *et al.* (eds), *Méthodologie de la recherche en sciences de gestion: Réussir son mémoire ou sa thèse*, Pearson, Paris, 2008.

[KÄH 13] KÄHKÖNEN A.-K., LINTUKANGAS K., HALLIKAS J., "Buyer dependency on suppliers in value-creative supplier relationships", *22nd Annual IPSERA Conference*, 2013.

[KAL 02] KALE P., DYER J.H., SINGH H., "Alliance capability, stock market response, and long-term alliance success: The role of the alliance function", *Strategic Management Journal*, vol. 23, pp. 747–767, 2002.

[KAL 07] KALE P., SINGH H., "Building firm capabilities through learning: The role of the alliance learning process in alliance capability and firm-level alliance success", *Strategic Management Journal*, vol. 28, pp. 981–1000, 2007.

[KAL 09] KALE P., SINGH H., "Managing strategic alliances: What do we know now, and where do we go from here?", *The Academy of Management Perspectives*, vol. 23, pp. 45–62, 2009.

[KAP 13] KAPOOR R., LEE J.M., "Cooperating and competing in ecosystems: how organizational forms shape new technology investments", *Strategic Management Jounral*, vol. 34, no. 3, pp. 274–295, 2013.

[KES 98] KESSELER A., "The creative supplier a new model for strategy, innovation, and customer relationships in concurrent design and engineering processes: The case of the automotive industry", *CRG Ecole Polytechnique*, Paris, 1998.

[KOS 09] KOSKINEN K.U., MÄKINEN S., "Role of boundary objects in negotiations of project contracts", *International Journal of Project Management*, vol. 27, pp. 31–38, 2009.

[KRA 83] KRALJIC P., "Purchasing must become supply management", *Harvard Business Review*, vol. 61, pp. 109–117, 1983.

[LAK 01] LAKEMOND N., Managing across organisations: Intra- and inter-organisational aspects of supplier involvement in product development projects, Thesis, 2001.

[LAK 06a] LAKEMOND N., BERGGREN C., "Co-locating NPD? The need for combining project focus and organizational integration", *Technovation*, vol. 26, pp. 807–819, 2006.

[LAK 06b] LAKEMOND N., BERGGREN C., VAN WEELE A., "Coordinating supplier involvement in product development projects: A differentiated coordination typology", *R & D Management*, vol. 36, pp. 55–66, 2006.

[LAK 14] LAKEMOND N., "Collaborative product development with suppliers – An evolutionary perspective", *Annual IPSERA Conference*, vol. 14, 2014.

[LAR 80] LARZELERE R.E., HUSTON T.L., "The dyadic trust scale: Toward understanding interpersonal trust in close relationships", *Journal of Marriage and Family*, vol. 42, pp. 595–604, 1980.

[LAS 00] LASBORDES V., *Les contrats déséquilibrés*, Presses universitaires d'Aix-Marseille, Aix-en-Provence, 2000.

[LED 10] LE DAIN M.-A., CALVI R., CHERITI S., "Developing an approach for design-or-buy-design decision-making", *Journal of Purchasing and Supply Management*, vol. 16, pp. 77–87, 2010.

[LED 11] LE DAIN M.-A., CALVI R., CHERITI S., "Measuring supplier performance in collaborative design: Proposition of a framework", *R & D Management*, vol. 41, pp. 61–79, 2011.

[LED 14] LE DAIN M.-A., MERMINOD V., "Le partage de connaissances dans un projet de codéveloppement", *Revue française de gestion*, no. 239, pp. 121–142, 2014.

[LEM 06] LE MASSON P., WEIL B., HATCHUEL A., *Les processus d'innovation: Conception innovante et croissance des entreprises*, Hermes-Lavoisier, 2006.

[LEE 12] LEE C.-J., JOHNSEN R.E., "Asymmetric customer–supplier relationship development in Taiwanese electronics firms", *Industrial Marketing Management*, vol. 41, pp. 692–705, 2012.

[LIN 11] LINDEGAARD S., *Making Open Innovation Work*, CreateSpace Independent Publishing Platform, North Charleston, 2011.

[LÔ 16] LÔ A., SACÉPÉ K., "Innovation Labs 2037 : Back to the future!", *Entreprendre & Innover*, pp. 62–69, 2016.

[LOI 11] LOILIER T., TELLIER A., "Que faire du modèle de l'innovation ouverte?", *Revue française de gestion*, no. 210, pp. 69–85, 2011.

[LUS 96] LUSCH R.F., BROWN J.R., "Interdependency, contracting, and relational behavior in marketing channels", *Journal of Marketing*, vol. 60, pp. 19–38, 1996.

[LUZ 15] LUZZINI D., AMANN M., CANIATO F. *et al.*, "The path of innovation: Purchasing and supplier involvement into new product development", *Industrial Marketing Management*, pp. 109–120, 2015.

[MAC 80] MACNEIL I.R., "Power, contract, and the economic model", *Journal of Economic Issues*, vol. 14, pp. 909–923,1980.

[MAI 17] MAIER M.A., RÜCK P., BREM A., "How to integrate suppliers into the innovation process? An explorative case of champion formalization in the purchasing department in times of open innovation", *International Journal of Innovation and Technology Management*, vol. 14, p. 1750036, 2017.

[MAN 09] MANIAK R., Les processus de co-innovation – Caractérisation, Evaluation et Management : Le cas de l'industrie automobile, Ecole Polytechnique X, 2009.

[MAR 93] MARCH J.G., SIMON H.A., *Organizations*, Wiley, 1993.

[MCA 95] McAllister D.J., "Affect- and cognition-based trust as foundations for interpersonal cooperation in organizations", *The Academy of Management Journal*, vol. 38, pp. 24–59, 1995.

[MEL 13] Melander L., Lopez-Vega H., "Impact of technological uncertainty in supplier selcction for NPD collaborations: Literature review and a case study", *International Journal of Technology Intelligence and Planning*, vol. 9, pp. 323–339, 2013.

[MEL 14] Melander L., Lakemond N., "Variation of purchasing's involvement: Case studies of supplier collaborations in new product development", *International Journal of Procurement Management*, vol. 7, pp. 103–118, 2014.

[MEL 15] Melander L., Lakemond N., "Governance of supplier collaboration in technologically uncertain NPD projects", *Industrial Marketing Management*, pp. 116–127, 2015.

[MER 13] Merminod N., Poissonnier H., "Implementing collaborative relationships with suppliers: Evolutions required of purchasing firms", *22nd Annual IPSERA Conference*, pp. 311–324, 2013.

[MID 93] Midler C., *L'Auto qui n'existait pas: Management des projets et transformation de l'entreprise*, Dunod, Paris, 1993.

[MID 96] Midler C., "Modèles gestionnaires et régulations économiques de la conception", *Cahiers du Centre de Recherche en Gestion*, pp. 11–26, 1996.

[MID 97] Midler C., Garel G., Kesseler A., "Le co-développement, définitions, enjeux et problèmes", *Education permanente*, vol. 95–108, 1997.

[MID 98] Midler C., "Évolution des modèles d'organisation et régulations économiques de la conception", *Problèmes économiques*, 1998.

[MID 00] Midler C., *Les partenariats interentreprises en conception: Pourquoi? Comment?*, vol. 54, 2000.

[MID 07] Midler C., Maniak R., Beaume R., "Du co-développement à la co-innovation – Analyse empirique des coopérations verticales en conception innovante", *15th GERPISA International Colloquium*, 2007.

[MIK 18] Mikkelsen O.S., Johnsen T.E., "Purchasing involvement in technologically uncertain new product development projects: Challenges and implications", *Journal of Purchasing and Supply Management*, 2018.

[MON 00] Monczka R.M., *New Product Development: Strategies for Supplier Integration*, ASQ Quality Press, 2000.

[MON 17] MONTEIRO F., BIRKINSHAW J., "The external knowledge sourcing process in multinational corporations", *Strategic Management Journal*, vol. 38, pp. 342–362, 2017.

[MOR 94] MORGAN R.M., HUNT S.D., "The commitment-trust theory of relationship marketing", *The Journal of Marketing*, vol. 58, pp. 20–38, 1994.

[MOR 11] MORTARA L., MINSHALL T., "How do large multinational companies implement open innovation?", *Technovation*, vol. 31, pp. 586–597, 2011.

[NOG 02] NOGATCHEWSKY G., "Typologie de controle interorganisationnel: Le cas de la relation entre un équipementier automobile et ses fournisseurs", *XIème conférence de l'AIMS*, 2002.

[NON 95] NONAKA I., TAKEUCHI H., *The Knowledge Creating Firm*, Oxford University Press, New York, 1995.

[NON 03] NONAKA I., TOYAMA R., "The knowledge-creating theory revisited: Knowledge creation as a synthesizing process", *Knowledge Management Research and Practice*, vol. 1, nos 2–10, 2003.

[NOO 97] NOOTEBOOM B., BERGER H., NOORDERHAVEN N.G., "Effects of trust and governance on relational risk", *The Academy of Management Journal*, vol. 40, pp. 308–338, 1997.

[NOO 99a] NOOTEBOOM B., "The triangle: Roles of the go-between", in LEENDERS R.T.A.J., GABBAY S.M. (eds), *Corporate Social Capital and Liability*, Springer, 1999.

[NOO 99b] NOOTEBOOM B., "Innovation and inter-firm linkages: New implications for policy", *Research Policy*, vol. 28, pp. 793–805, 1999.

[NOO 07] NOOTEBOOM B., VANHAVERBEKE W., DUYSTERS G. *et al.*, "Optimal cognitive distance and absorptive capacity", *Research Policy*, vol. 36, pp. 1016–1034, 2007.

[OUC 80] OUCHI W.G., "Markets, bureaucracies, and clans", *Administrative Science Quarterly*, vol. 25, pp. 129–141, 1980.

[PAR 93] PARKHE A., "Strategic alliance structuring: A game theoretic and transaction cost examination of interfirm cooperation", *Academy of Management Journal*, vol. 36, pp. 794–829, 1993.

[PAR 98] PARKHE A., "Building trust in international alliances", *Journal of World Business*, vol. 33, pp. 417–437, 1998.

[PAT 17] PATRUCCO A.S., RONCHI S., LUZZINI D., "Achieving innovation through supplier collaboration – The role of the purchasing interface", *Business Process Management Journal*, vol. 23, no. 1, 2017.

[PER 13] PERSONNIER H., LE DAIN M.-A., SCHIELE H. *et al.*, "Failures in collaborative design with suppliers: Impact analysis on project innovation", *22nd IPSERA Annual Conference*, pp. 1075–1095, 2013.

[PET 05] PETERSEN K.J., HANDFIELD R.B., RAGATZ G.L., "Supplier integration into new product development: Coordinating product, process and supply chain design", *Journal of Operations Management*, vol. 23, pp. 371–388, 2005.

[POO 04] POOLE M.S., VAN DE VEN A.H., *Handbook of Organizational Change and Innovation*, Oxford University Press, Oxford, 2004.

[PRO 89] PROVAN K.G., SKINNER S.J., "Interorganizational dependence and control as predictors of opportunism in dealer-supplier relations", *The Academy of Management Journal*, vol. 32, pp. 202–212, 1989.

[PUR 09] PURANAM P., VANNESTE B., "Trust and governance: Untangling a tangled Web", *The Academy of Management Review ARCHIVE*, vol. 34, pp. 11–31, 2009.

[RET 11] RETAILLEAU B., KIRSCH A.-R., FAUCHEUX M. *et al.*, Les entreprises de taille intermédiaire au coeur d'une nouvelle dynamique de croissance, Report, 2011.

[REU 07] REUER J.J., ARIÑO A., "Strategic alliance contracts: Dimensions and determinants of contractual complexity", *Strategic Management Journal*, vol. 28, pp. 313–330, 2007.

[RIN 92] RING P.S., VAN DE VEN A.H., "Structuring cooperative relationships between organizations", *Strategic Management Journal*, vol. 13, pp. 483–498, 1992.

[RIN 94] RING P.S., VAN DE VEN A.H., "Developmental processes of cooperative interorganizational relationships", *The Academy of Management Review*, vol. 19, pp. 90–118, 1994.

[ROB 12] ROBERTS N., GALLUCH P.S., DINGER M. *et al.*, "Absorptive capacity and information systems research: Review, synthesis, and directions for future research", *MIS Quarterly*, vol. 36, no. 2, pp. 625–648, 2012.

[ROS 14] ROSELL D.T., LAKEMOND N., WASTI S.N., "Integrating knowledge with suppliers at the R & D-manufacturing interface", *Journal of Manufacturing Technology Management*, vol. 25, pp. 240–257, 2014.

[ROY 04] ROY S., SIVAKUMAR K., WILKINSON I.F., "Innovation generation in supply chain relationships: A conceptual model and research propositions", *Journal of the Academy of Marketing Science*, vol. 32, pp. 61–79, 2004.

[SAD 00] SADEH A., DVIR D., SHENHAR A., "The role of contract type in the success of R & D defense projects under increasing uncertainty", *Project Management Journal*, vol. 31, pp. 14–22, 2000.

[SÄF 14] SÄFSTEN K., JOHANSSON G., LAKEMOND N. *et al.*, "Interface challenges and managerial issues in the industrial innovation process", *Journal of Manufacturing Technology Management*, vol. 25, pp. 218–239, 2014.

[SAK 92] SAKO M., *Prices, Quality, and Trust: Inter-firm Relations in Britain and Japan*, Cambridge University Press, 1992.

[SAP 04] SAPIENZA H.J., PARHANKANGAS A., AUTIO E., "Knowledge relatedness and post-spin-off growth", *Journal of Business Venturing*, vol. 19, pp. 809–829, 2004.

[SCH 06] SCHIELE H., "How to distinguish innovative suppliers? Identifying innovative suppliers as new task for purchasing", *Industrial Marketing Management*, vol. 35, pp. 925–935, 2006.

[SCH 10] SCHIELE H., "Early supplier integration: The dual role of purchasing in new product development", *R & D Management*, vol. 40, pp. 138–153, 2010.

[SCH 12] SCHIELE H., "Accessing supplier innovation by being their preferred customer", *Research Technology Management*, vol. 55, pp. 44–50, 2012.

[SEG 03] SEGRESTIN B., La gestion des partenariats d'exploration : Spécificités, crises et formes de rationalisation, École Nationale Supérieure des Mines de Paris, 2003.

[SER 15] SERVAJEAN-HILST R., Approche relationnelle de la coopération verticale d'innovation – Facteurs de performance de la coopération client-fournisseur en innovation, Ecole polytechnique, Paris, 2015.

[SER 17] SERVAJEAN-HILST R., "Cinq problématiques clef pour impliquer les fournisseurs dans un projet d'innovation", *Entreprendre & Innover*, vol. 4, pp. 29–39, 2017.

[SER 18a] SERVAJEAN-HILST R., "Necessary governing practices for the success (and failure) of client-supplier innovation cooperation", in MOREIRA A.C., FERREIRA L.M.D.F., ZIMERMAN R.A. (eds), *Innovation and Supply Chain Management: Relationship, Collaboration and Strategies*, Springer, 2018.

[SER 18b] SERVAJEAN-HILST R., CALVI R., "Shades of the innovation-purchasing function – The missing link of open innovation", *International Journal of Innovation Management*, vol. 22, p. 1850008, 2018.

[SIV 00] SIVADAS E., DWYER F.R., "An examination of organizational factors influencing new product success in internal and alliance-based processes", *Journal of Marketing*, vol. 64, pp. 31–49, 2000.

[SLU 11] SLUYTS K., MATTHYSSENS P., MARTENS R. *et al.*, "Building capabilities to manage strategic alliances", *Industrial Marketing Management*, vol. 40, pp. 875–886, 2011.

[SOB 02] SOBRERO M., ROBERTS E.B., "Strategic management of supplier–manufacturer relations in new product development", *Research Policy*, vol. 31, pp. 159–182, 2002.

[TAK 86] TAKEUCHI H., NONAKA I., "The new new product development game", *Harvard Business Review*, vol. 64, pp. 137–146, 1986.

[TAK 01] TAKEISHI A., "Bridging inter- and intra-firm boundaries: Management of supplier involvement in automobile product development", *Strategic Management Journal*, vol. 22, pp. 403–433, 2001.

[TEE 07] TEECE D.J., "Explicating dynamic capabilities: The nature and microfoundations of (sustainable) enterprise performance", *Strategic Management Journal*, vol. 28, pp. 1319–1350, 2007.

[THI 59] THIBAUT J.W., KELLEY H.H., *The Social Psychology of Groups*, Wiley, Oxford, 1959.

[THI 99] THIETART R.-A., *Méthodes de recherche en management*, Dunod, Paris, 1999.

[TRA 09] TRAITLER H., SAGUY L.S., "Creating successful innovation partnerships", *Food Technology*, vol. 63, pp. 23–35, 2009.

[TRE 98] TRENT R.J., MONCZKA R.M., "Purchasing and supply management: Trends and changes throughout the 1990s", *Journal of Supply Chain Management*, vol. 34, pp. 2–11, 1998.

[TSA 98] TSAI W., GHOSHAL S., "Social capital and value creation: The role of intrafirm networks", *Academy of Management Journal*, vol. 41, pp. 464–476, 1998.

[USU 00] USUNIER J.-C., *Confiance et performance: Un essai de management comparé France-Allemagne*, Vuibert, 2000.

[UTT 75] UTTERBACK J.M., ABERNATHY W.J., "A dynamic model of process and product innovation", *Omega*, vol. 3, pp. 639–656, 1975.

[VAN 95] VAN DE VEN A.H., POOLE M.S., "Explaining development and change in organizations", *The Academy of Management Review*, vol. 20, pp. 510–540, 1995.

[VAN 01] VAN ECHTELT F.E.A., WYNSTRA F., "Managing supplier integration into product development: A literature review and conceptual model", *The Future of Innovation Studies*, 2001.

[VAN 08] VAN ECHTELT F.E.A., WYNSTRA F., VAN WEELE A.J. *et al.*, "Managing supplier involvement in new product development: A multiple-case study", *Journal of Product Innovation Management*, vol. 25, pp. 180–201, 2008.

[VAN 09] VAN DE VRANDE V., VANHAVERBEKE W., DUYSTERS G., "External technology sourcing: The effect of uncertainty on governance mode choice", *Journal of Business Venturing*, vol. 24, pp. 62–80, 2009.

[VAN 14] VANHAVERBEKE W., DU J., LETEN B. *et al.*, "Exploring open innovation at the level of R & D projects", *New Frontiers in Open Innovation*, Oxford, 2014.

[VOL 10] VOLOT J.-C., Le dispositif juridique concernant les relations interentreprises et la sous-traitance – Rapport du Médiateur des relations inter-entreprises industrielles et de la sous-traitance, Report, 2010.

[WES 14] WEST J., BOGERS M., "Leveraging external sources of innovation: A review of research on open innovation", *Journal of Product Innovation Management*, vol. 31, pp. 814–831, 2014.

[WHE 92] WHEELWRIGHT S.C., CLARK K.B., "Creating project plans to focus product development", *Harvard Business Review*, 1992.

[WIL 75] WILLIAMSON O.E., *Markets and Hierarchies: Antitrust Analysis and Implications*, Free Press New York, 1975.

[WIL 85] WILLIAMSON O.E., *The Economic Institutions of Capitalism: Firms, Markets, Relational Contracting*, vol. 472, Free Press New York, 1985.

[WIL 93] WILLIAMSON O.E., "Calculativeness, trust, and economic organization", *Journal of Law and Economics*, vol. 36, pp. 453–86, 1993.

[WYN 98] WYNSTRA F., VAN WEELE A., AXELSSON B., *The Role of Purchasing in Product Development: A Summary of Four Years Research*, vol. 545, 1998.

[WYN 99] WYNSTRA F., VAN WEELE A., AXELSSON B., "Purchasing involvement in product development: A framework", *European Journal of Purchasing and Supply Management*, vol. 5, pp. 129–141, 1999.

[WYN 03] WYNSTRA F., WEGGEMAN M., VAN WEELE A., "Exploring purchasing integration in product development", *Industrial Marketing Management*, vol. 32, pp. 69–83, 2003.

[YLI 14] YLIMÄKI J., "A dynamic model of supplier–customer product development collaboration strategies", *Industrial Marketing Management*, vol. 43, pp. 996–1004, 2014.

[ZAH 98] ZAHEER A., MCEVILY B., PERRONE V., "Does trust matter? Exploring the effects of interorganizational and interpersonal trust on performance", *Organization Science*, vol. 9, pp. 141–159, 1998.

[ZAH 02] ZAHRA S.A., GEORGE G., "Absorptive capacity: A review, reconceptualization, and extension", *Academy of Management Review*, vol. 27, pp. 185–203, 2002.

[ZHA 11] ZHANG C., VISWANATHAN S., HENKE JR. J.W., "The boundary spanning capabilities of purchasing agents in buyer–supplier trust development", *Journal of Operations Management*, vol. 29, pp. 318–328, 2011.

[ZOL 02] ZOLLO M., REUER J.J., SINGH H., "Interorganizational routines and performance in strategic alliances", *Organization Science*, vol. 13, pp. 701–713, 2002.

Index

Other titles from

ISTE

in

Innovation, Entrepreneurship and Management

2019

FRIMOUSSE Soufyane
Innovation and Agility in the Digital Age
(Human Resources Management Set – Volume 2)

HÉRAUD Jean-Alain, KERR Fiona, BURGER-HELMCHEN Thierry
Creative Management of Complex Systems
(Smart Innovation Set – Volume 19)

LATOUCHE Pascal
Open Innovation: Corporate Incubator
(Innovation and Technology Set – Volume 7)

LEHMANN Paul-Jacques
The Future of the Euro Currency

MILLOT Michel
Embarrassment of Product Choices 2: Towards a Society of Well-being

PRIOLON Joël
Financial Markets for Commodities

ROGER Alain, VINOT Didier
Skills Management: New Applications, New Questions
(Human Resources Management Set – Volume 1)

SKIADAS Christos H., BOZEMAN James R.
Data Analysis and Applications 1: Clustering and Regression, Modeling-estimating, Forecasting and Data Mining
(Big Data, Artificial Intelligence and Data Analysis Set – Volume 2)
Data Analysis and Applications 2: Utilization of Results in Europe and Other Topics
(Big Data, Artificial Intelligence and Data Analysis Set – Volume 3)

2018

BURKHARDT Kirsten
Private Equity Firms: Their Role in the Formation of Strategic Alliances

CALLENS Stéphane
Creative Globalization
(Smart Innovation Set – Volume 16)

CASADELLA Vanessa
Innovation Systems in Emerging Economies: MINT – Mexico, Indonesia, Nigeria, Turkey
(Smart Innovation Set – Volume 18)

CHOUTEAU Marianne, FOREST Joëlle, NGUYEN Céline
Science, Technology and Innovation Culture
(Innovation in Engineering and Technology Set – Volume 3)

CORLOSQUET-HABART Marine, JANSSEN Jacques
Big Data for Insurance Companies
(Big Data, Artificial Intelligence and Data Analysis Set – Volume 1)

CROS Françoise
Innovation and Society
(Smart Innovation Set – Volume 15)

DEBREF Romain
Environmental Innovation and Ecodesign: Certainties and Controversies
(Smart Innovation Set – Volume 17)

DOMINGUEZ Noémie
SME Internationalization Strategies: Innovation to Conquer New Markets

ERMINE Jean-Louis
Knowledge Management: The Creative Loop
(Innovation and Technology Set – Volume 5)

GILBERT Patrick, BOBADILLA Natalia, GASTALDI Lise,
LE BOULAIRE Martine, LELEBINA Olga
Innovation, Research and Development Management

IBRAHIMI Mohammed
Mergers & Acquisitions: Theory, Strategy, Finance

LEMAÎTRE Denis
Training Engineers for Innovation

LÉVY Aldo, BEN BOUHENI Faten, AMMI Chantal
Financial Management: USGAAP and IFRS Standards
(Innovation and Technology Set – Volume 6)

MILLOT Michel
Embarrassment of Product Choices 1: How to Consume Differently

PANSERA Mario, OWEN Richard
Innovation and Development: The Politics at the Bottom of the Pyramid
(Innovation and Responsibility Set – Volume 2)

RICHEZ Yves
Corporate Talent Detection and Development

SACHETTI Philippe, ZUPPINGER Thibaud
New Technologies and Branding
(Innovation and Technology Set – Volume 4)

SAMIER Henri
Intuition, Creativity, Innovation

TEMPLE Ludovic, COMPAORÉ SAWADOGO Eveline M.F.W.
Innovation Processes in Agro-Ecological Transitions in Developing Countries
(Innovation in Engineering and Technology Set – Volume 2)

UZUNIDIS Dimitri
Collective Innovation Processes: Principles and Practices
(Innovation in Engineering and Technology Set – Volume 4)

VAN HOOREBEKE Delphine
The Management of Living Beings or Emo-management

2017

AÏT-EL-HADJ Smaïl
The Ongoing Technological System
(Smart Innovation Set – Volume 11)

BAUDRY Marc, DUMONT Béatrice
Patents: Prompting or Restricting Innovation?
(Smart Innovation Set – Volume 12)

BÉRARD Céline, TEYSSIER Christine
Risk Management: Lever for SME Development and Stakeholder Value Creation

CHALENÇON Ludivine
Location Strategies and Value Creation of International Mergers and Acquisitions

CHAUVEL Danièle, BORZILLO Stefano
The Innovative Company: An Ill-defined Object
(Innovation Between Risk and Reward Set – Volume 1)

CORSI Patrick
Going Past Limits To Growth

D'ANDRIA Aude, GABARRET Inés
Building 21st Century Entrepreneurship
(Innovation and Technology Set – Volume 2)

DAIDJ Nabyla
Cooperation, Coopetition and Innovation
(Innovation and Technology Set – Volume 3)

FERNEZ-WALCH Sandrine
The Multiple Facets of Innovation Project Management
(Innovation between Risk and Reward Set – Volume 4)

FOREST Joëlle
Creative Rationality and Innovation
(Smart Innovation Set – Volume 14)

GUILHON Bernard
Innovation and Production Ecosystems
(Innovation between Risk and Reward Set – Volume 2)

HAMMOUDI Abdelhakim, DAIDJ Nabyla
Game Theory Approach to Managerial Strategies and Value Creation
(Diverse and Global Perspectives on Value Creation Set – Volume 3)

LALLEMENT Rémi
Intellectual Property and Innovation Protection: New Practices and New Policy Issues
(Innovation between Risk and Reward Set – Volume 3)

LAPERCHE Blandine
Enterprise Knowledge Capital
(Smart Innovation Set – Volume 13)

LEBERT Didier, EL YOUNSI Hafida
International Specialization Dynamics
(Smart Innovation Set – Volume 9)

MAESSCHALCK Marc
Reflexive Governance for Research and Innovative Knowledge
(Responsible Research and Innovation Set – Volume 6)

MASSOTTE Pierre
Ethics in Social Networking and Business 1: Theory, Practice and Current Recommendations
Ethics in Social Networking and Business 2: The Future and Changing Paradigms

MASSOTTE Pierre, CORSI Patrick
Smart Decisions in Complex Systems

MEDINA Mercedes, HERRERO Mónica, URGELLÉS Alicia
Current and Emerging Issues in the Audiovisual Industry
(Diverse and Global Perspectives on Value Creation Set – Volume 1)

MICHAUD Thomas
Innovation, Between Science and Science Fiction
(Smart Innovation Set – Volume 10)

PELLÉ Sophie
Business, Innovation and Responsibility
(Responsible Research and Innovation Set – Volume 7)

SAVIGNAC Emmanuelle
The Gamification of Work: The Use of Games in the Workplace

SUGAHARA Satoshi, DAIDJ Nabyla, USHIO Sumitaka
Value Creation in Management Accounting and Strategic Management: An Integrated Approach
(Diverse and Global Perspectives on Value Creation Set –Volume 2)

UZUNIDIS Dimitri, SAULAIS Pierre
Innovation Engines: Entrepreneurs and Enterprises in a Turbulent World
(Innovation in Engineering and Technology Set – Volume 1)

2016

BARBAROUX Pierre, ATTOUR Amel, SCHENK Eric
Knowledge Management and Innovation
(Smart Innovation Set – Volume 6)

BEN BOUHENI Faten, AMMI Chantal, LEVY Aldo
Banking Governance, Performance And Risk-Taking: Conventional Banks Vs Islamic Banks

BOUTILLIER Sophie, CARRÉ Denis, LEVRATTO Nadine
Entrepreneurial Ecosystems (Smart Innovation Set – Volume 2)

BOUTILLIER Sophie, UZUNIDIS Dimitri
The Entrepreneur (Smart Innovation Set – Volume 8)

BOUVARD Patricia, SUZANNE Hervé
Collective Intelligence Development in Business

GALLAUD Delphine, LAPERCHE Blandine
Circular Economy, Industrial Ecology and Short Supply Chains (Smart Innovation Set – Volume 4)

GUERRIER Claudine
Security and Privacy in the Digital Era (Innovation and Technology Set – Volume 1)

MEGHOUAR Hicham
Corporate Takeover Targets

MONINO Jean-Louis, SEDKAOUI Soraya
Big Data, Open Data and Data Development (Smart Innovation Set – Volume 3)

MOREL Laure, LE ROUX Serge
Fab Labs: Innovative User (Smart Innovation Set – Volume 5)

PICARD Fabienne, TANGUY Corinne
Innovations and Techno-ecological Transition (Smart Innovation Set – Volume 7)

2015

CASADELLA Vanessa, LIU Zeting, DIMITRI Uzunidis
Innovation Capabilities and Economic Development in Open Economies (Smart Innovation Set – Volume 1)

CORSI Patrick, MORIN Dominique
Sequencing Apple's DNA

CORSI Patrick, NEAU Erwan
Innovation Capability Maturity Model

FAIVRE-TAVIGNOT Bénédicte
Social Business and Base of the Pyramid

GODÉ Cécile
Team Coordination in Extreme Environments

MAILLARD Pierre
Competitive Quality and Innovation

MASSOTTE Pierre, CORSI Patrick
Operationalizing Sustainability

MASSOTTE Pierre, CORSI Patrick
Sustainability Calling

2014

DUBÉ Jean, LEGROS Diègo
Spatial Econometrics Using Microdata

LESCA Humbert, LESCA Nicolas
Strategic Decisions and Weak Signals

2013

HABART-CORLOSQUET Marine, JANSSEN Jacques, MANCA Raimondo
VaR Methodology for Non-Gaussian Finance

2012

DAL PONT Jean-Pierre
Process Engineering and Industrial Management

MAILLARD Pierre
Competitive Quality Strategies

POMEROL Jean-Charles
Decision-Making and Action

SZYLAR Christian
UCITS Handbook

2011

LESCA Nicolas
Environmental Scanning and Sustainable Development

LESCA Nicolas, LESCA Humbert
Weak Signals for Strategic Intelligence: Anticipation Tool for Managers

MERCIER-LAURENT Eunika
Innovation Ecosystems

2010

SZYLAR Christian
Risk Management under UCITS III/IV

2009

COHEN Corine
Business Intelligence

ZANINETTI Jean-Marc
Sustainable Development in the USA

2008

CORSI Patrick, DULIEU Mike
The Marketing of Technology Intensive Products and Services

DZEVER Sam, JAUSSAUD Jacques, ANDREOSSO Bernadette
Evolving Corporate Structures and Cultures in Asia: Impact of Globalization

2007

AMMI Chantal
Global Consumer Behavior

2006

BOUGHZALA Imed, ERMINE Jean-Louis
Trends in Enterprise Knowledge Management

CORSI Patrick *et al.*
Innovation Engineering: the Power of Intangible Networks